BASIC SKILLS

Arithmetic

John Deft

JOHN MURRAY

Students' Book   ISBN 0 7195 4349 5
Teachers' Resource Book   ISBN 0 7195 4356 8

Other books in the Basic Skills Series

English by Paul Groves and Nigel Grimshaw with *Teachers' Resource Book*
Electronics by Tom Duncan
Health, Hygiene and Safety by Di Barton and Wilf Stout
Science by Peter Leckstein
Geography by Sally Naish and Katherine Goldsmith with *Teachers' Resource Book*

Also available
Lifestart by John Anderson .

Acknowledgements
The map on page 152 is based upon the Ordnance Survey Landranger map number 140 with the permission of the Controller of Her Majesty's Stationery Office, Crown Copyright reserved. The timetable on page 116 is reproduced by kind permission of British Rail. Cartoons by John Erasmus. Line diagrams by RDL Artset.

© John Deft 1988

First published 1988
by John Murray (Publishers) Ltd
50 Albemarle Street, London W1X 4BD

Reprinted 1993, 1994, 1995, 1997, 1999, 2002

Printed and bound in Great Britain by
Athenaeum Press Ltd, Gateshead, Tyne & Wear

British Library Cataloguing in Publication Data

Deft, John
  Basic skills : arithmetic.
  Students' book
  1. Arithmetic—1961—
  I. Title
  513   QA107
  ISBN 0-7195-4349-5

Contents

Preface

In life, skills in basic arithmetic are often useful and sometimes vital: even the best pocket calculator in the world is no good if you do not know which buttons to press. Many employers in particular look for arithmetical ability in their workforce.

This book is designed to provide a course in the basic arithmetic needed in work and in everyday life. To make the best use of arithmetical skills, it is essential to understand the simple ideas involved. All the topics are therefore developed in easy-to-follow steps and followed by many examples for practice and assessment.

The content of this book is based on the AEB Basic Test in Arithmetic, but also covers the needs of several other pre-vocational courses.

The accompanying *Teachers' Resource Book* is designed to allow for photocopying, and contains further exercises, revision papers, answers to questions, and notes on some of the practical work and investigations.

Unit 1 | Whole Numbers

Why do we count in tens? Because we have ten fingers and thumbs! These are sometimes called digits. The figures 0, 1, 2, 3, 4, 5 . . . to 9 are also called digits.

We can show numbers of any size with these ten digits in our **decimal place-value** system. **Decimal** means numbered by tens. A digit has different meanings according to where it is placed in a number. Its **place** shows its **value**.

3 in 358 is 3 hundreds
3 in 137 is 3 tens

A _____ Write down what the digit (figure) 7 stands for in each of these numbers.

1 175	**3** 736	**5** 3758	**7** 9570	**9** 37 144
2 217	**4** 1027	**6** 7061	**8** 10 407	**10** 73 348

B _____ We can write numbers in figures or words. e.g. 1748 can be written as 'one thousand seven hundred and forty eight'.

Now write down the numbers in Exercise A in words.

The figure 0

The figure 0 is very important to make our system work. We use it to show that there are no units, or no tens, or no thousands. You cannot write three hundred and seven as 37! Think of it as 3 hundreds, no tens and seven units and write 307.

C _____ Write these numbers in figures.

1 two hundred and thirty-six
2 four hundred and seventy
3 six hundred and one
4 one thousand two hundred
5 one thousand six hundred and forty-two
6 two thousand five hundred and three
7 four thousand and seventy-five
8 six thousand and twenty
9 seven thousand and eight
10 nine thousand three hundred and eighty

Putting figures in the right order

With more than one digit you can make different numbers. 2 and 1 can make 21 or 12. This picture of six sacks shows the numbers you can make with the digits 3, 4 and 7.

Which number is smallest? Which number is biggest?

D _____ 1 What is the largest number you can make using the digits 3, 6, 2 and 5 (once each)?
2 What is the smallest whole number you can make using the digits 2, 8 and 5?
3 List all the three-digit numbers you can make using the digits 4, 1 and 9.
4 What is the largest three-digit number that has three different digits?
5 What is the smallest whole number that has three different digits?

_____ 7 _____

Counting

Counting is something that you probably find quite easy, but you still need to be careful at times. For instance, the next whole number after 6399 is 6400 – a few people get confused by this.

E ——— Check your counting with these questions.

1 What is the next whole number after 1339?
2 What number is one more than 1699?
3 What number is one less than 2000?
4 What is the whole number immediately before 2150?
5 What number is one more than 2409?

Egyptian Numbers

The ancient Egyptians used a number system with different symbols for units, tens, and so on. Three of the symbols they used were:

| for 1

∩   for 10

℮   for 100

and they simply repeated each symbol as often as necessary. For example, to write 317 the Egyptians would have written:

℮ ℮ ℮ ∩ |||

F _____ Write these Egyptian numbers in our way.

1 |||

2 ||||

3 ∩|

4 ∩∩

5 ∩∩∩ ||| |||

6 ∩∩∩ ∩∩ |

7 ℮||

8 ℮∩∩∩

9 ℮∩ |||| ||||

10 ℮℮℮

G _____ Write these numbers in the Egyptian way.

1 5	**3** 13	**5** 29	**7** 40	**9** 200
2 9	**4** 25	**6** 34	**8** 110	**10** 703

H _____ Which do you think is the better number system: the Egyptian one or the modern one?

Negative numbers

Sometimes we have to deal with **negative** as well as **positive** numbers, for example, when we talk about temperatures below freezing, or bank accounts 'in the red', or heights below sea level in low-lying country.

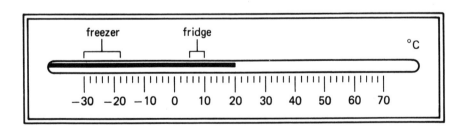

The easiest way to handle these numbers is to draw a number line like the one below and use this for counting. (We have shown positive numbers to the right and negative to the left, but many people like to show positive upwards and negative downwards.)

I ———— Use the number line to help you answer these questions.

1 What number is one more than −5?

2 What number is seven more than −3?

3 What number is three less than 2?

4 What number is two less than −4?

5 How much warmer is 1°C than −5°C?

6 How much colder is −10°C than −8°C?

7 Which of these numbers is the largest: −3, 5, −7?

8 Which of these numbers is the smallest: −2, 8, −9?

9 The top of Mount Everest is 8 km above sea level; the bottom of the Mindanao Trench is 14 km below sea level. How much higher is Mount Everest than the Mindanao Trench?

10 A man has a bank balance of −£50 (that is, he already owes the bank £50). If he withdraws £10 more, what is his balance now?

Addition and Subtraction

Practice with small numbers

A ──────── Try these few easy questions, working them out in your head.

$6+9=?$

1 Add six and nine.
2 Take three from fourteen.
3 Add twelve to seventeen.
4 Subtract seven from nineteen.
5 What is the sum of eleven and thirteen?
6 Increase eighteen by five.
7 Find the difference between six and twenty.
8 What is ten more than eighteen?
9 Decrease twenty-six by fifteen.
10 Take twelve from thirty.

Adding larger numbers

When you add larger numbers, you will probably want to write them down.

$385+17+2046=?$

$$\begin{array}{r} 385 \\ 17 \\ +2046 \\ \hline 2448 \end{array}$$

Make sure that:

all the units figures are in line (5, 7 and 6)

all the tens figures are in line (8, 1 and 4) and so on ...

Then add each column in turn, **starting at the right**.

The answer is shown. Ask your teacher if you do not understand it.

Now try these.

1 804 + 27
2 995 + 1056
3 78 + 164
4 590 + 455
5 1972 + 337

6 449 + 801 + 188
7 45 + 228 + 520
8 488 + 16 + 361
9 7 + 178 + 39
10 1057 + 44 + 221

Subtraction

164 − 23 = ?

$$\begin{array}{r} 1\ 6\ 4 \\ -\ \ 2\ 3 \\ \hline 1\ 4\ 1 \end{array}$$

152 − 39 = ?

$$\begin{array}{r} 1\ 5\ 2 \\ -\ \ 3\ 9 \\ \hline 1\ 1\ 3 \end{array}$$

Subtraction is set out like addition. The first sum is easy.

The second sum is more difficult because we can't 'do' 2 − 9, without using negative numbers. So we change the 5 tens into 4 tens and 10 units, to give us 2 + 10 = 12 units. 12 − 9 = 3.

In the next column, 4 − 3 = 1, and in the left-hand column 1 − 0 = 1 so the answer is 113. (If you use another method of 'borrowing' and 'paying back', go on using it if you get the right answers.)

C

1 884 − 213
2 316 − 195
3 850 − 226
4 1690 − 448
5 705 − 33

6 186 − 45
7 2106 − 383
8 1015 − 788
9 500 − 245
10 347 − 88

D

1 1352 + 467
2 289 − 105
3 1582 + 2954
4 351 − 227
5 1046 − 319

6 58 + 227 + 163
7 603 − 257
8 583 − 21
9 64 + 105 − 7
10 8 + 5 + 17 + 48

Magic squares

9	10	5
4	8	12
11	6	7

A **magic square** is an arrangement of numbers (all different) in a square so that each row, each column, and each diagonal adds up to the same total. Here is a magic square:

You can check that $9 + 10 + 5$, and $4 + 8 + 12$, and $11 + 6 + 7$, and $9 + 4 + 11$, and $10 + 8 + 6$, and $5 + 12 + 7$, and $9 + 8 + 7$, and $5 + 8 + 11$ each add up to 24.

E _____ Copy these magic squares and fill in the missing numbers:

1

	2	
	6	
3	10	

2

4	18	
	10	
		16

3

		10
	1	25
		4

4

2	9	16	7
12		1	13
15			10

5

1			8
	14	11	2
		4	
10		6	15

F _____ Now make your own magic square! Use the numbers 1 to 9 once each. You could cut out nine small pieces of paper and write one number on each piece. Then you can move these around to try to make the square. This is hard, and will probably take quite a long time. (*Hint:* The 'magic total' is 15.)

Draw your square in your book.

5	10	3	16
4	15	6	9
14	1	12	7
11	8	13	2

G Look at the magic square above. Each row, each column, and each diagonal adds up to 34 – you can check that if you want – but there are lots of other patterns too. For example, the four numbers in the top left corner (5 + 10 + 4 + 15) give 34; so do the four corners (5 + 16 + 11 + 2).

Try to find as many sets as you can of four numbers adding up to 34 – you should be able to find at least twenty sets.

(This is sometimes called a **diabolical** magic square!)

H The page below comes from a milkman's delivery book, and shows how many pints of milk he delivered to each house in a street on each day last week.

HOUSE NUMBER	1	3	5	7	8	6	4	2	TOTAL
SUNDAY	2	4	1	3	6	2	1	2	
MONDAY	3	4	1	3	7	2	0	2	
TUESDAY	2	5	1	3	6	2	1	2	
WEDNESDAY	2	4	1	3	6	2	0	2	
THURSDAY	3	4	1	3	6	3	1	2	
FRIDAY	2	5	1	3	3	2	0	2	1
SATURDAY	2	5	1	3	0	2	1	2	
TOTAL									

Work out

1 How many pints altogether the milkman delivered each day.

2 How many pints each house received during the week.

Check that when you add the totals you have listed for **1** and for **2** that they give the same **grand total**.

Unit 3 | Multiplication

Multiplying and adding

Multiplying whole numbers is the same as **repeated addition**.

$4 \times 6 = 6 + 6 + 6 + 6 = 24$

As you probably know, two numbers give the same answer when multiplied the other way round. (We say multiplication is **commutative**.) So

$6 \times 4 = 4 + 4 + 4 + 4 + 4 + 4 = 24$

This rule applies when **any** two numbers are multiplied together.

A ———— Write each of these multiplications as an addition.

1 3×5	**6** 4×9
2 5×3	**7** 2×8
3 2×7	**8** 5×10
4 7×2	**9** 6×1
5 4×8	**10** 5×5

B ———— Without working out the answers, write down **another** multiplication that has the same answer as each of those below. For example, $3 \times 9 = 9 \times 3$.

1 5×7	**6** 8×2
2 4×10	**7** 9×5
3 11×6	**8** 4×30
4 7×8	**9** 117×32
5 3×16	**10** $A \times B$

Know your tables!

To do multiplication easily and well, it is a great help to know your tables. If you don't, now is a good time to learn. If you do, try the next exercise. Write answers only.

C ——— Work as fast as you can and time yourself. You should finish in 2 minutes, but add ten seconds for each wrong answer.

1 3 × 5	**6** 4 × 8	**11** 4 × 9	**16** 6 × 9
2 6 × 2	**7** 8 × 6	**12** 8 × 5	**17** 3 × 10
3 2 × 9	**8** 4 × 4	**13** 7 × 7	**18** 7 × 9
4 5 × 10	**9** 1 × 7	**14** 10 × 6	**19** 8 × 8
5 3 × 7	**10** 9 × 5	**15** 9 × 8	**20** 9 × 3

Multiplying larger numbers: the distributive rule

The **distributive rule** allows us to break numbers into bits and multiply them separately.

So 4 × 83 can be split up into (4 × 80) + (4 × 3)

$$= 320 + 12$$
$$= 332$$

```
  8 3
×   4
3 3 2
    1
```

Most people write this out as shown here.

First multiply 4 × 3 to give 12. Write down 2 units and carry 1 ten.

Then 4 × 80 gives 32 tens, plus one ten carried, which gives 33 tens altogether.

D ——— Work out the following by any suitable method (but not with a calculator). Remember that you can write the numbers the other way round if it helps.

1 5 × 79	**6** 4 × 28
2 7 × 36	**7** 5 × 167
3 3 × 88	**8** 275 × 2
4 56 × 4	**9** 6 × 182
5 8 × 49	**10** 3 × 406

Multiplying larger numbers: the associative rule

Multiplying by ten is easy.

For example $93 \times 10 = 930$

This can be split up into

$$3 \times 10 = 30 = 3 \text{ tens}$$
$$\text{and } 90 \times 10 = 900 = 9 \text{ hundreds}$$

$$58 \times 40 = (58 \times 4) \times 10$$
$$= 232 \times 10$$
$$= 2320$$

When you multiply by ten, the figures stay the same, but **move one column to the left**. We can now use the **associative** rule, which allows us to multiply by (say) 40 in two stages. First we multiply by 4, then by 10.

E ———— Work out the following.

1 32×20 **3** 50×48 **5** 103×60 **7** 245×30 **9** 300×82

2 73×40 **4** 30×146 **6** 70×61 **8** 100×58 **10** 156×200

Combining methods

By putting the methods together, you can multiply much larger numbers. Suppose you want to multiply 136 by 86.

The stages are:

$$136 \times 86 = (100 \times 86) + (30 \times 86) + (6 \times 86)$$
$$= 8600 \quad + 2580 \quad + 516$$
$$= 11\,696$$

Or:

$$136 \times 86 = (136 \times 80) + (136 \times 6)$$
$$= 10\,880 \quad + 816$$
$$= 11\,696$$

Often, this is set out as a **long multiplication**. The answer is reached just as in the method above.

$$136 \times 86 = ?$$

$$\begin{array}{r} 136 \\ \times\ 86 \\ \hline 10880 \\ 816 \\ \hline 11696 \end{array}$$

F ———— Work out each of these by any suitable method (but not with a calculator).

1 35×62 **3** 82×27 **5** 62×44 **7** 175×31 **9** 491×125

2 37×23 **4** 12×75 **6** 48×105 **8** 76×284 **10** 106×285

Unit 4 Division

What is division?

You can think of division in two ways.

Either as the inverse (opposite) of multiplication, so $126 \div 7$ means 'how many 7's make 126?'.
The answer is 18, because $18 \times 7 = 126$.

Or it is a problem of sharing.
So $126 \div 7$ means: 'if there are 126 objects shared between 7 people, how many will each person get?'
The answer is 18, again.

$126 \div 7$ is **not** the same as $7 \div 126$, so division is **not** commutative. You must get the numbers the right way round.

The distributive and associative rules do not work for division either.

Setting out division

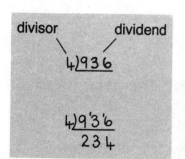

There are several good methods of setting out division. The following way works well, but if you have been taught another way and can get right answers, stick to it.

Suppose you have to divide 936 by 4. This may be written as $936 \div 4$ or 936/4 or $\frac{936}{4}$.

Write the numbers with a division bracket.
The *dividend* (number to be divided) goes inside the bracket.
The *divisor* (number doing the dividing) goes outside the bracket.
Now start to divide at the left (the hundreds figure). $9 \div 4$ is 2 with remainder 1 (because $2 \times 4 = 8$).
Write 2 in the hundreds column of the answer.
1 hundred is 10 tens, so there are 13 tens altogether.

$13 \div 4$ is 3, again with 1 remaining. Write 3 in the tens column of the answer.

1 ten is 10 units, so there are 16 units altogether.

$16 \div 4$ is 4 and so the answer is 234.

(Some people write the answer above the division bracket, but the working is just the same.)

A _____ Work out the following divisions:

1 $576 \div 4$	**6** $144 \div 9$
2 $655 \div 5$	**7** $296 \div 8$
3 $732 \div 3$	**8** $612 \div 3$
4 $889 \div 7$	**9** $847 \div 7$
5 $96 \div 6$	**10** $1356 \div 6$

'Long' division

In long division, the divisor is more than 10.

To divide 4128 by 24:

Set out as before. Start with the thousands.

4 will not divide by 24 so this leaves us with 41 hundreds.

$41 \div 24 = 1$ with 17 remaining. Write the one in the hundreds column of the answer.

We now have 172 tens.

$172 \div 24 = 7$ ($7 \times 24 = 168$) with 4 remaining.

This gives 48 units.

$48 \div 24 = 2$ and so the answer is 172.

$$24 \overline{)41\overset{17}{2}\overset{4}{8}}$$
$$172$$

B _____ Work out the following divisions in your own way (not with a calculator).

1 $742 \div 14$	**6** $855 \div 45$
2 $729 \div 27$	**7** $176 \div 11$
3 $6336 \div 44$	**8** $6405 \div 105$
4 $775 \div 25$	**9** $8778 \div 42$
5 $3165 \div 15$	**10** $4830 \div 21$

Problems with remainders

$319 \div 4 = ?$

$$4\overline{)319}$$
$$79 \text{ rem } 3$$

All the questions you have done gave exact answers, but this does not always happen. In this question, $319 \div 4$, there is a remainder of 3 after dividing the units. One way of dealing with this is to leave the remainder. Here, the answer is 79 rem 3. Some of the next set of questions have remainders.

C —————— Work out the following, writing any remainders in the answers.

1 $658 \div 5$

2 $295 \div 8$

3 $407 \div 4$

4 $330 \div 5$

5 $535 \div 7$

6 $3815 \div 17$

7 $453 \div 25$

8 $748 \div 28$

9 $1297 \div 32$

10 $2951 \div 50$

Will there be a remainder?

It is useful to know in advance whether a division will have a remainder. A simple test for dividing by 2 is:

even numbers (ending in 0, 2, 4, 6, 8) divide by 2

odd numbers (ending 1, 3, 5, 7, 9) will not

D _____ Without doing any of the divisions, say whether each of these numbers will divide exactly by 2.

35	64	53	88	92
36	49	60	71	32
107	372	933	804	776
704	160	388	279	105

E _____ Try to write down for yourself a rule that says which numbers will divide exactly by 5. (*Hint:* Look at the last figure.)

Dividing by 3

There is a simple rule to find if a number will divide exactly by 3.
Add up the digits and if the total will divide by 3, the number will also divide by 3.
Thus 417 will divide by 3 because $4 + 1 + 7 = 12$, which divides by 3 (check $417 \div 3 = 139$).
616 will not divide by 3 because $6 + 1 + 6 = 13$ which will not divide by 3 (check $616 \div 3 = 205$ rem 1).

F _____ Which of these numbers will divide exactly by 3?

48	57	64	77	95
32	45	73	78	89
102	135	188	247	594

G _____ The smallest positive number that divides exactly by 2 *and* by 3 is 6. The smallest positive number that divides exactly by 2, by 3 *and* by 4 is 12.

What is the smallest positive number that divides exactly by 2, by 3, by 4 *and* by 5? Will it divide by 6 too?

Unit 5 Mixed Operations

$$3+4\times2=?$$
$$3+4=7 \qquad 7\times2=14$$
$$4\times2=8 \qquad 3+8=11$$

$$3+(4\times2)=11$$
$$(3+4)\times2=14$$

How do you deal with this calculation?

Which of the two ways shown is right? We could do the sum either way and get two different answers. To avoid confusion, we use **brackets**. So that:

$3+4\times2$ can be written as $3+(4\times2)=11$.

But if the sum is written:

$(3+4)\times2$, the answer is 14.

Mathematicians always do mixed calculations in the same order:

1 Work out anything in brackets

2 Do the multiplication and divisions, left to right.

3 Do the additions and subtractions, left to right.

These five examples show how the rules work.

a $3+4\times2=3+8=11$

b $(7+2)\times(3-1)=9\times2=18$

c $7+2\times3-1=7+6-1=12$

d $7+2\times(3-1)=7+2\times2=7+4=11$

e $(7+2)\times3-1=9\times3-1=27-1=26$

A ———— Work out the following, being sure to do things in the right order.

1 $8-(3+2)$

2 $(8-3)+2$

3 $5\times(6\div3)$

4 $(3\times7)+3$

5 $3\times(7+3)$

6 $3\times7+3$

7 $10-5-3$

8 $10-(5-3)$

9 $8+(2\times5)$

10 $(8+2)\times5$

11 $7\times2+10$

12 $3\times5+4\times2$

13 $6-5+2\times3$

14 $8\times4-2+6$

15 $12\div3\times2$

16 $12\div(3\times2)$

17 $6+3\times4-5$

18 $9-2\times4+6$

19 $16-3\times(4-1)$

20 $(16-3)\times4-1$

Averages

This shows the heights (**H**) and weights (**W**) of four girls.

H (cm)	160	165	168	171
W (kg)	50	45	55	60

Their **average** height is 166 cm. This is calculated by adding all their heights and dividing by the number of girls.

$(160 + 165 + 168 + 171) \div 4$

$= 664 \div 4 = 166$ cm (add first, then divide).

What is the average weight of the girls?

In the same way, we can calculate the average score for each batsman.

Gotham
0, 58, 40, 36, 166, 4, 60, 84.

For Gotham:
Total = 448
Average = $448 \div 8 = 56$

Bower
40, 60, 20, 88, 102, 80, 50, 40.

For Bower:
Total = 480
Average = $480 \div 8 = 60$

There are several kinds of average but the kind in these examples is the most common one. It is called the **mean**.

B _____ Work out the average of each of these sets.

1. Test marks: 6, 7, 4, 8, 10.
2. Egg weights: 25 g, 25 g, 30 g, 28 g, 30 g, 24 g.
3. Occupants of cars: 1, 3, 1, 1, 2, 5, 1, 3, 2, 1.
4. Matches in a box: 46, 48, 47, 45, 50, 46.
5. Children in each family: 2, 4, 0, 2, 1, 5, 6, 4.
6. Exam results: 70, 80, 40, 45, 35.
7. Shoe sizes: 4, 4, 4, 7, 5, 2, 9, $5\frac{1}{2}$, 6, $3\frac{1}{2}$.
8. Length of nails: 1″, 1″, 1″, 2″, 2″, 4″, 4″, 6″, 6″.
9. Weekly wages: £50, £50, £50, £300, £50.
10. Hours worked: 10, 12, 8, 12.

C _____ A *palindrome* is a number (or a word or phrase) that reads the same backwards as it does forwards; for example, 3443 and 15251 are both palindromes.

Start with any two-digit number that is not already a palindrome; write it down both forwards and backwards, and add these two together.
If the answer is not a palindrome, write it down forwards and backwards, and so on until you get a palindromic answer. For example, if you start with 68:

$68 + 86 = 154$
$154 + 451 = 605$
$605 + 506 = 1111$, which is a palindrome.

This example took three steps to give a palindrome.

Can you find starting numbers that take **a** two steps, **b** four steps, **c** six steps, **d** more than six steps, to make a palindrome?

Unit 6 | Number Problems

Calculators are quick and accurate. They are a marvellous help when faced with arithmetic that will take a long time with pencil and paper.

But a calculator cannot tell you what **operation** is needed. *You* must know when to add, subtract, divide or multiply.

A calculator cannot tell you whether the answer it gives is sensible. *You* must check this.

A ———— Work out these problems in your head.

1. If eggs are sold in boxes of six, how many eggs will there be in four boxes?

2. In a class there are 11 boys and 14 girls. How many pupils are there in the class altogether?

3. How many ice-creams at 30p each could you buy for 90p?

4. From a packet of 12 pencils, four pencils are missing; how many pencils are there left in the packet?

5. A bus starts its journey with seven passengers. At the next stop, three passengers get off and four others get on. How many passengers are there on the bus now?

6. If three girls share 27 toffees equally between them, how many toffees does each girl get?

7. This week's Number 1 hit record has moved up from 7th place last week. How many places has it moved up?

8. A darts player throws three darts and scores 1, 20 and double 5. What is his score for these three darts?

9. How much would it cost to buy three tins of baked beans at 16p each?

10. A snooker player pots red, then black, then red, then pink, then red. If red is worth 1 point, pink 6 points, and black 7 points, what is her total score for this break?

B _____ Work out these problems, using pencil and paper if you wish.

1. Susan has collected 778 foreign stamps. How many more must she collect before she has a thousand?

2. A car goes 35 miles on each gallon of petrol; how far will it go on seven gallons?

3. A tea merchant has 180 kg of Tips, 593 kg of Blend, and 227 kg of Best. How much tea has he altogether?

4. A long-distance walker travels 343 km in seven days. How far (on average) does he go each day?

5. Each week you should be at school for five days. If there are 38 weeks in the school year, how many days each year should you be at school?

6. Four boys are playing marbles. Matthew has 26 marbles, Mark has 105, Luke has 33 and John has just 8. If they put all their marbles together and then shared them equally between them, how many would each boy get?

7. In a scientific experiment Peggy mixes 35 g of sulphur with 41 g of iron. What is the total weight of the mixture?

8. 624 people go to a concert, but 47 of them get bored and leave at the interval. How many people are still at the concert after the interval?

9. In the election for the school council, Wayne got 112 votes, Saroop got 145 and David got 38. How many votes were cast altogether?

10. How many pencils at 7p each could I buy for 30p?

C _____ Work out these problems, using a calculator if it helps.

1. The *Daily Liar* sells 395 100 copies each day; how many copies will be sold in a week (6 days)?

2. If pencils are sold in packets of 12, how many packets would you need to buy to get 600 pencils?

3. When metal is heated it expands. A certain metal bar measured 1142 mm when cool and 1155 mm when hot; how much longer had it become?

4. A running track is 400 m long. How many laps of the track would be run in the 10 000 m race?

5. Jason has a lot of books, each 3 cm thick. How many of these books could he fit onto a shelf 135 cm long?

6 Ten light bulbs are tested until they go out, and last for 893, 610, 1024, 1335, 936, 998, 1046, 1292, 877 and 799 hours. What was their average life?

7 In an election, Smith got 21 793 votes and Jones got 16 772 votes. What was Smith's majority? (That is, how many *more* votes did Smith get than Jones?)

8 Donna buys three boxes each containing 120 Smarties; Jeffrey buys ten tubes each containing 38 Smarties. Which of them has more Smarties, and by how many?

9 Queen Victoria came to the throne in 1837 and died in 1901; how long did her reign last?

10 A school has 1102 pupils; at the end of the year 237 pupils leave the fifth year but 201 others join the first year. How many pupils has the school in the new year?

D ———— These problems are harder, so you need to think carefully to get the right answers. Use a calculator when you need to.

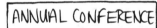

1 A straight fence is made from ten posts spaced three metres apart. How long is the fence from end to end?

2 A bottle and a cork together cost 24p. If the bottle costs 20p more than the cork, how much does the cork cost?

3 A plank 12 feet long is cut into two-foot lengths. If each cut is made separately, how many cuts are needed?

4 At a committee meeting seven members voted 'Yes' to a particular motion and four voted 'No'. How many of the Yeses would have to change their vote in order for the Noes to win?

5 At a party political conference there were 120 delegates present. If each delegate shook hands (once) with every other delegate, how many handshakes took place altogether?

6 At the same conference, one speech was very boring. Of the 120 delegates, 27 fell asleep, 51 went out for a drink, and all but six of the rest read their newspapers. How many delegates were still listening to the speech?

7 230 children are going on a school outing. If each coach will seat 49 children, how many coaches will be needed altogether?

8 How many whole numbers are there *between* 87 and 99?

9 If it takes four minutes to boil an egg, how long does it take to boil three eggs?

10 If seven is two less than one too many, how many is enough?

E _____ By writing the figures 1, 2, 3, 4, 5 (once each) in any order in the five boxes below, what is the largest answer you can get? (Use a calculator to help you explore the possibilities.)

$$\square \ \square \ \square \ \times \ \square \ \square \ = \ ?$$

F _____ When Karl Gauss (who later became a very famous mathematician) was still at school, his teacher decided to keep the class busy by getting them to add together all the numbers from 1 to 100 inclusive. In less than a minute, young Karl was at the teacher's desk with the right answer.

Can you work out how he might have done it, remembering of course that he did not have a calculator?

Factors and Multiples

$$10 \div 10 = 1$$
$$10 \div 5 = 2$$
$$10 \div 4 = 2 \text{ rem } 2$$
$$10 \div 3 = 3 \text{ rem } 1$$
$$10 \div 2 = 5$$
$$10 \div 1 = 10$$

The **factors** of a number are the whole numbers that divide into it exactly. 10, 5, 2 and 1 are factors of 10. 4 and 3 are **not** factors of 10.

To find factors it is sometimes easier to ask 'How do I get the number as an answer to a multiplication?' Here are the four ways of doing this for 10.

$$10 \times 1 = 10$$
$$1 \times 10 = 10$$
$$2 \times 5 = 10$$
$$5 \times 2 = 10$$

The four numbers involved are 10, 5, 2 and 1. So these are the factors of ten.

A _____ Write down the factors of these numbers.

1 6	**6** 7	**11** 12	**16** 60
2 8	**7** 11	**12** 24	**17** 32
3 15	**8** 25	**13** 20	**18** 72
4 16	**9** 18	**14** 45	**19** 29
5 9	**10** 13	**15** 48	**20** 30

Highest common factor

20	30
1	1
2	2
4	3
5	5
10	6
20	10
	15
	30

Here are the factors of 20 and 30. The **biggest** number that is in **both** lists is 10.

So the **highest common factor** (H.C.F.) of 20 and 30 is 10. In the same way the H.C.F. of 9 and 12 is 3, the H.C.F. of 6 and 8 is 2.

B _____ What is the highest common factor of these pairs of numbers?

1	6 and 10	**6**	12 and 15
2	10 and 15	**7**	30 and 50
3	18 and 30	**8**	20 and 24
4	12 and 24	**9**	15 and 16
5	20 and 21	**10**	20 and 36

Prime Numbers

Every whole number more than 1 has at least **two** factors, because 1 is a factor of every number, and every number is a factor of itself. There are some numbers, though, which have **only** these two factors: the factors of 13, for instance, are 1 and 13. Numbers like this are called **prime numbers**. So 13 is a prime number, because it has only two factors; but 14 is not prime because it has factors 1, 2, 7 and 14.

C _____ Write 'prime' or 'not prime' for each of these numbers.

1 7		**3** 9		**5** 11		**7** 17		**9** 21	
2 8		**4** 10		**6** 15		**8** 19		**10** 24	

D _____ Make a list of all the prime numbers under 50, starting with 2. (*Hint:* there are fifteen of them altogether.)

E _____ Many people have tried to invent rules for finding prime numbers. One rule that has been suggested is 'start with any number, multiply it by six, and take away one from the answer.'

Try this rule several times, starting with a different number each time. Is the answer at the end always a prime number?

Multiples of numbers

$1 \times 10 = 10$
$2 \times 10 = 20$
$3 \times 10 = 30$
$4 \times 10 = 40$

This is the ten times table. The numbers 10, 20, 30, 40 . . . (on the right) are **multiples** of 10. Of course this list can go on for ever (is said to be **infinite**) and you cannot write all the multiples of any number.

F _____ Write down the **first six** multiples of each of these numbers.

1 2	**3** 7	**5** 8	**7** 6
2 5	**4** 4	**6** 3	**8** 9

G _____ Write down the **first three** multiples of each of these numbers:

1 12	**5** 20	**9** 22
2 24	**6** 45	**10** 50
3 48	**7** 25	**11** 21
4 15	**8** 36	**12** 16

Lowest common multiple

	20	30
$\times 1 =$	20	30
$\times 2 =$	40	60
$\times 3 =$	60	90
$\times 4 =$	80	120
$\times 5 =$	100	150

Here are some multiples of 20 and 30. Look at the lists.

The **smallest** multiple in **both** lists is 60. We call this the **lowest common multiple** (L.C.M.).

H _____ Find the lowest common multiple of each of these pairs of numbers.

1 4 and 6	**6** 10 and 15
2 4 and 8	**7** 12 and 36
3 6 and 8	**8** 4 and 18
4 5 and 6	**9** 6 and 10
5 3 and 15	**10** 12 and 20

I _____ Start with any number between 1 and 100.

> If it is *even*, halve it (that is, divide it by 2);
> If it is odd, multiply it by 3 and add 1.

Look at your answer.

> If it is even, halve it;
> If it is odd, multiply it by 3 and add 1.

Go on doing this until you get 1 as the answer.

For example, if you start with 13 you get the chain

$$13 \rightarrow 40 \rightarrow 20 \rightarrow 10 \rightarrow 5 \rightarrow 16 \rightarrow 8 \rightarrow 4 \rightarrow 2 \rightarrow 1.$$

This chain contains ten numbers; some chains will be longer or shorter.

Try it for several different numbers; what number (between 1 and 100) gives the longest chain?

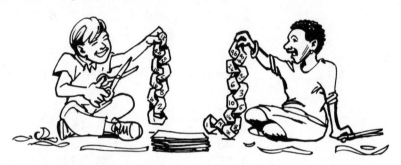

J _____ The **Goldbach Conjecture** is a rule in advanced mathematics. It says that every even number from 6 upwards can be got by adding two prime numbers. Using your list of numbers from exercise **D**, write down the prime numbers needed to give these answers.

1 12	**3** 20	**5** 30	**7** 38	**9** 48
2 16	**4** 24	**6** 36	**8** 44	**10** 50

Unit 8 | Powers and Roots

Powers

3^1 is 3
3^2 is $3 \times 3 = 9$
3^3 is $3 \times 3 \times 3 = 27$
3^4 is $3 \times 3 \times 3 \times 3 = 81$

Sometimes you have to multiply a number by itself several times. The table shows how this is done.

$3 \times 3 \times 3 \times 3$ is written 3^4: we say 'three to the power four' or 'three to the fourth power'.

Notice that the answer is 81 and don't confuse it with 3×4 (answer 12) or 4^3 (answer 64).

A ——— Write these out in full:

1 2^5	**3** 3^5	**5** 5^6	**7** 9^4	**9** 1^7
2 4^2	**4** 8^2	**6** 6^3	**8** 10^3	**10** 12^2

Working out the answers

$3^4 = 3 \times 3 \times 3 \times 3$
$= 9 \times 3 \times 3$
$= 27 \times 3$
$= 81$

Here is a simple way. You multiply two numbers at a time which is easy. You can also use a calculator.

B ——— Write out in full and then work out the answers.

1 4^3	**3** 5^2	**5** 2^7	**7** 3^3	**9** 6^3
2 2^5	**4** 8^2	**6** 7^2	**8** 10^4	**10** 1^8

Squares and cubes

Two kinds of power happen so often that they have their own names.

A number to the power of 2 is called a **square** number.

A number to the power of 3 is called a **cube**.

So 4^2 is 'four to the power two' or 'four squared' and 5^3 is 'five to the power three' or 'five cubed'.

C ——— Read these numbers saying 'squared' or 'cubed' as appropriate.

1 6^2	**3** 10^2	**5** 8^2	**7** 5^2	**9** 1^3
2 7^3	**4** 4^3	**6** 14^3	**8** 11^3	**10** 3^3

D ——— Write out as multiplications, and then work out the answers.

1 five squared
2 seven squared
3 four cubed
4 eight cubed
5 nine squared

6 ten squared
7 two cubed
8 one squared
9 three cubed
10 six cubed

Roots

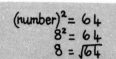

$(number)^2 = 64$
$8^2 = 64$
$8 = \sqrt{64}$

Here a number has been multiplied by itself to give 64.

You know that $8 \times 8 = 64$ so the number is 8. We say that 8 is the **square root** of 64.

In the same way $3 \times 3 = 9$ so 3 is the **square root** of 9.
So we can write them:
$$8 = \sqrt{64}$$
$$3 = \sqrt{9}$$

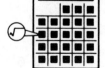

Using a calculator with a $\sqrt{}$ button, check the answer by pressing 64 and *then* pressing $\sqrt{}$.

E _____ Work out the following square roots (your tables will help you).

1 $\sqrt{25}$ 3 $\sqrt{49}$ 5 $\sqrt{36}$ 7 $\sqrt{4}$ 9 $\sqrt{100}$

2 $\sqrt{81}$ 4 $\sqrt{9}$ 6 $\sqrt{16}$ 8 $\sqrt{0}$ 10 $\sqrt{1}$

Use a calculator to check your answers.

Powers to describe large numbers

Writing very big numbers can be made a lot easier using powers of 10.

For example 90 000 000 can be written as 9×10^7 because
$$9 \times 10^7 = 9 \times 10 \times 10 \times 10 \times \times 10 \times 10 \times 10 \times 10$$
$$= 90 \times 10 \times 10 \times 10 \times 10 \times 10 \times 10$$
$$= 900 \times 10 \times 10 \times 10 \times 10 \times 10$$

and so on.

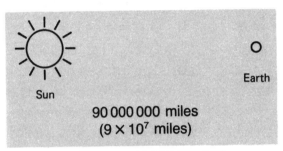

Sun

90 000 000 miles
(9×10^7 miles)

Earth

In the same way $2 \times 10^{13} = 20\,000\,000\,000\,000$.

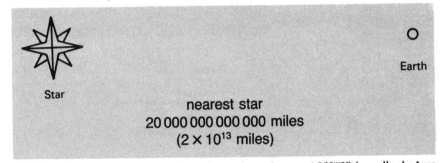

Star

nearest star
20 000 000 000 000 miles
(2×10^{13} miles)

Earth

This scientific way of writing numbers (number $\times 10^{power}$) is called **standard form**. Some calculators show numbers this way, e.g. 9   07, when they are too big to display in full.

F _____ Multiply out these numbers.

1 3×10^2 6 5×10^3

2 7×10^4 7 1×10^6

3 6×10^3 8 2×10^4

4 3×10^5 9 4×10^7

5 9×10^3 10 1×10^{10}

Unit 9 | Decimal Fractions

Place value

The decimal system works with whole numbers. It also works with parts of numbers (fractions).

HTU t h

4 1 6 · 8 2

H One hundred can be split into ten tens

T One ten can be split into ten units

U One unit can be split into ten tenths

t One tenth can be split into ten hundredths

h One hundredth can be split into ten thousandths.

It is vital to know which column is which. To make this clear we put a small dot, **a decimal point**, after the units column and before the tenths.

'four hundred and sixteen point eight two'

So 416·82 is said as 'four hundred and sixteen **point** eight two'.

A _____ Write these in figures, using a decimal point.

1 three tens, five units and six tenths

2 four tens, two units, eight tenths and seven hundredths

3 two hundreds, three tens, six units and nine tenths

4 five units, eight tenths and one hundredth

5 seven units, no tenths and six hundredths

6 eight tens, nine units, no tenths and five hundredths

7 six hundreds, no tens, three units, and five tenths

8 no units, six tenths, and four hundredths

9 three units, four tenths, no hundredths and seven thousandths

10 no units, no tenths, and one hundredth

B _____ Say which is the **units** figure in each of the following:

1 35·4   **3** 0·51   **5** 38·02   **7** 189·3   **9** 481·0

2 6·72   **4** 178·6   **6** 20·59   **8** 3·219   **10** 697·25

C _____ The figure 6 in 31·62 stands for 'six tenths'. What does the 6 stand for in each of these numbers?

1 36·27   **3** 58·62   **5** 0·693   **7** 20·61   **9** 3·846

2 61·8   **4** 2·06   **6** 9·865   **8** 0·06   **10** 6·928

Comparing numbers

Which of these is larger: **A** 0·63 or **B** 0·27?

The number **A** is 6 tenths and 3 hundredths (63 hundredths).

The number **B** is 2 tenths and 7 hundredths (27 hundredths).

You can see that the first number is bigger because it contains more tenths.

Use the same reasoning with these numbers: **C** 0·06 and **D** 0·5.

The number **C** is 6 hundredths.

The number **D** is 5 tenths. Which is larger?

Tenths are much bigger than hundredths, so **D** is larger.

In these cases the important figure is the first or **significant** figure. In the examples above the significant figures are:

A 6 = 6 tenths   **B** 2 = 2 tenths

C 6 = 6 hundredths   **D** 5 = 5 tenths or 50 hundredths.

D _____ Compare each pair of numbers and write down which is **larger**.

1 0·48 and 0·71   **6** 0·05 and 0·31

2 0·8 and 0·36   **7** 0·2 and 0·49

3 0·42 and 0·1   **8** 0·04 and 0·083

4 0·57 and 0·7   **9** 0·93 and 1·26

5 0·265 and 0·19   **10** 1·69 and 1·34

E Choose the **smallest** of each of these sets.

1 0·35, 0·27, 0·49
2 0·82, 0·61, 0·58
3 0·6, 0·48, 0·73
4 0·062, 0·31, 0·075
5 0·48, 0·41, 0·69

6 0·07, 0·7, 7·0
7 0·61, 0·68, 0·6
8 0·59, 1·38, 0·8
9 1·5, 1·33, 1·82
10 0·2, 0·25, 0·237

Accuracy

0·2	0 units and 2 tenths
0·20	0 units and 2 tenths and 0 hundredths
0·200	0 units and 2 tenths and 0 hundredths and 0 thousandths

All these numbers are equal, so why are they written differently?

It is because we want to show different **levels of accuracy**.

0·2 is accurate to 1 decimal place

0·20 is accurate to 2 decimal places

0·200 is accurate to 3 decimal places

F To how many decimal places is each of these numbers given?

1 0·54	3 0·9	5 0·70	7 3·61	9 0·057
2 0·540	4 0·07	6 0·7	8 4·9	10 7·500

Suppose you do a calculation with money on a calculator. 7 people share £25. How much does each get? My calculator gives the answer shown. Of course our money is measured to the nearest penny, so £3.57 is the only sensible answer. One person will get an extra penny!

Three places	Two places
0·476	0·48
0·294	0·29
0·357	0·36
0·353	0·35
0·485	0·49

Four places	Two places
0·3682	0·37
0·1628	0·16

We say that we have written the answer to **two decimal places**.

In the same way 0·472 written to two decimal places is 0·47. But what will 0·479 be to two places?

0·479 = 0·47 + 9 thousandths (nearly 10 thousandths or 1 hundredth)

So 0·479 = 0·48 to two decimal places.

Look at this table. Can you see the simple rule for converting to two decimal places?

The rule is: **look at the third decimal place**. If it's less than 5 forget it. If it's 5 or more, add 1 to the second place.

Check with the list to see how the rule is applied.

G _____ Write each of these numbers to two decimal places.

1 0·354 **3** 0·186 **5** 4·1357 **7** 0·7016 **9** 7·398

2 1·278 **4** 0·354 **6** 3·2195 **8** 0·207 **10** 2·68

3·2$\overset{*}{7}$ to 1 place is 3·3

3·4$\overset{*}{7}$1 to 1 place is 3·5

3·4$\overset{*}{4}$9 to 2 places is 3·45

6·244$\overset{*}{8}$ to 3 places is 6·245

These examples show how you can use the same rule to shorten to any number of decimal places. If you want the number to *one* decimal place, it is the number in the *second* column (starred) that tells you whether to add an extra tenth. And, for three decimal places, it is the number in the *fourth* column. And so on.

H _____ Shorten each of these numbers as instructed.

1 0·271 to one decimal place

2 4·1832 to three dec. pl.

3 5·178 to two dec. pl.

4 0·027 to one dec. pl.

5 0·6125 to three dec. pl.

6 0·41 to one dec. pl.

7 2·96 to one dec. pl.

8 6·0581 to three dec. pl.

9 0·975 to one dec. pl.

10 3·14159 to four dec. pl.

Unit 10

Approximations and Errors

The method used at the end of Unit 9 can be applied to almost all numbers. For example: 473 to the *nearest hundred* is 500 (the first figure after the hundreds column is 7. This is more than 5 so add an extra hundred and get 500.). Common sense also tells you that 473 is nearer to 500 than 400.

Another example: 26·31 to the *nearest whole number* is 26.

A

1 Write 3417 to the nearest hundred.
2 Write 295 to the nearest hundred.
3 Write 437 to the nearest ten.
4 What is 31·8 to the nearest whole number?
5 What is 45·28 to the nearest whole number?
6 Write 1836 to the nearest thousand.
7 What is 3219 to the nearest ten?
8 What is 497 to the nearest hundred?
9 Write 39·72 to the nearest whole number.
10 What is 2385 to the nearest thousand?

Finding rough answers

Look at this calculation.

$72 \times 49 = ?$

You can get a rough answer in your head by saying '72 is just over 70, 49 is just under 50 and $70 \times 50 = 3500$'. So a rough answer is 3500. Check with your calculator. $72 \times 49 = 3528$, so the rough answer is close.

B —————— By approximating each number **to the nearest ten**, work out rough answers to each of these. (The first one is done for you.)

1 $21 \times 42 \approx 20 \times 40 = 800$　　　**6** $39 \times 52 =$

2 $29 \times 83 =$　　　**7** $11 \times 120 =$

3 $62 \times 37 =$　　　**8** $26 \times 37 =$

4 $98 \times 46 =$　　　**9** $34 \times 74 =$

5 $50 \times 68 =$　　　**10** $108 \times 31 =$

C —————— Use a calculator to work out actual answers to the questions in Exercise **B**. Compare these with your rough answers.

Rough answers are usually quite close, but there is almost always a **rounding error** when you make approximations.

D —————— A visitor to a museum asked the assistant how old a particular piece of Roman pottery was.
'Two thousand and one years old, miss,' said the assistant.
'How do you know?' asked the visitor.
'Well,' said the assistant, 'I started work here last year, and the curator told me then that the pot was two thousand years old.'
Why is the assistant being silly?

Significant figures

Here is another kind of approximation, based on **significant figures**.

36·027	Five significant figures
36·03	Four significant figures
36·0	Three significant figures
36	Two significant figures
40	One significant figure

Notice that in the last line we have to put a 0 in the empty units column (otherwise the number would just read 4), but this 0 is not counted as significant.

E ____

1 Write 27·46 correct to three significant figures.

2 What is 3·891 correct to two significant figures?

3 Write 372 to two significant figures.

4 Express 2950 correct to one significant figure.

5 What is 4·2557 to two significant figures?

6 Write 3·261 correct to three significant figures.

7 Write 0·438 to two significant figures. (Don't count the 0.)

8 Write 0·519 correct to one significant figure.

9 Correct 3·1874 to four significant figures.

10 Express 387 correct to one significant figure.

More rough answers

| 316 × 88 = ? |
| 300 × 90 = 27 000 |

What is 316 × 88?

You can get very close to the answer by writing each of the two numbers to 1 significant figure, and working out the answer in your head.

Use your calculator to check: 316 × 88 = 27 808, so the rough answer is quite close.

F ____

Write each number to one significant figure and then work out a rough answer in your head to each of the following:

1 32 × 106

2 213 × 51

3 49 × 72

4 92 × 34

5 143 × 195

6 4·25 × 9·86

7 38 × 2·86

8 137 × 5·23

9 1·83 × 3·15

10 7·41 × 16·85

G ____

Use a calculator to work out actual answers to the calculations in Exercise **F**; compare them with your rough answers.

Estimating and checking

$14.62 \times 9.61 =$

1404.982

but $15 \times 10 = 150$

Suppose you multiply 14·62 by 9·61 on a calculator. It is very easy to press the wrong button, or put the decimal point button down at the wrong time.

By **approximating**, you can see that the answer should be about 150. You can therefore see that the right answer is *not* the figure displayed. (Perhaps it should be 140·4982?)

Whenever you do a sum on a calculator, check that the answer is roughly the right size.

You won't spot **all** your mistakes by this sort of check, but you will certainly find some of them.

H _____ Here are twenty questions and their answers. Ten of the answers are right, but the other ten are very wrong. Use a **mental** check to mark each question right or wrong.

1 $394 \times 51 = 20\,094$

2 $325 \div 41 = 36$

3 $691 - 344 = 517$

4 $107 \times 48 = 5136$

5 $35 + 173 = 523$

6 $22 \times 817 = 17\,974$

7 $431 \times 728 = 3138$

8 $295 - 106 = 189$

9 $459 \div 17 = 27$

10 $176 \times 36 = 12\,672$

11 $278 + 195 = 354$

12 $29 \times 45 = 1305$

13 $352 \div 44 = 80$

14 $637 - 29 = 608$

15 $456 \div 38 = 12$

16 $26·2 \times 31·7 = 830·5$

17 $18·5 - 7·6 = 33·7$

18 $38·5 \div 6·3 = 6·1$

19 $12·38 + 9·42 = 116·62$

20 $45·32 \times 3·75 = 16·95$

Unit 11

Arithmetic with Decimals

Adding and subtracting

$$
\begin{array}{r}
38 \cdot 6 \\
+\ \ 5 \cdot 47 \\
\hline
\end{array}
$$

$$
\begin{array}{r}
20 \cdot 3 \\
14 \\
+\ \ 8 \cdot 56 \\
\hline
\end{array}
$$

We can add and subtract decimals in almost exactly the same way as whole numbers (Unit 2). The main rule to remember is:
All the units figures must be in line.

This is how you set out 38·6 + 5·47. Add in the usual way (starting at the right) and you get the answer 44·07.

Here is an addition with a mixture of whole numbers and decimals. 20·3 + 14 + 8·56. The units figures (0, 4 and 8) are in line. Add up as before and check that your answer is 42·86.

A ———— Add the following.

1 5·36 + 4·42
2 6·71 + 21·8
3 0·59 + 0·76
4 4·5 + 6·12
5 6·195 + 0·582
6 4·22 + 0·731
7 14·9 + 5·25
8 6·64 + 2·255
9 6·692 + 0·107
10 0·36 + 1·2

11 5·32 + 8
12 93·8 + 2·34
13 4·3 + 6·67
14 15 + 2·2
15 25·3 + 18
16 56 + 12·4 + 8·8
17 55·02 + 0·046 + 0·73
18 36·8 + 49 + 2·75
19 25 + 38 + 16·3
20 21·3 + 13 + 7·5

$$
\begin{array}{r}
39 \cdot 52 \\
-\ \ 4 \cdot 76 \\
\hline
\end{array}
$$

Subtracting is as easy as adding. Just put the units figures in line again. To work out 39·52 − 4·76, write the numbers as shown. Subtract as with whole numbers. Check your answer. Did you make it 34·76?

$$\begin{array}{r} 16{\cdot}5 \\ -\ 3{\cdot}29 \\ \hline \end{array} = \begin{array}{r} 16{\cdot}50 \\ -\ 3{\cdot}29 \\ \hline \end{array}$$

$$\begin{array}{r} 8 \\ -2{\cdot}73 \\ \hline \end{array} = \begin{array}{r} 8{\cdot}00 \\ -2{\cdot}73 \\ \hline \end{array}$$

Sometimes, one number has more decimal places than the other. Most people find it helpful to fill in with 0's for the empty space on the right hand side.

Here we see the way to set out
16·5 − 3·29   (answer 13·21)
8 − 2·73   (answer 5·27)

Remember the main rule about setting out:
Keep all the units figures in line.

KEEP ALL THE UNIT FIGURES IN LINE

B _____ Work out the following:

1 4·85 − 2·21	11 7·9 − 4·56
2 6·37 − 4·19	12 6 − 3·7
3 9·94 − 3·79	13 0·57 − 0·281
4 3·183 − 1·004	14 1·376 − 1·35
5 7·0 − 2·4	15 67 − 25·43
6 7·48 − 3·2	16 10 − 5·32
7 3·03 − 2·86	17 3·47 − 1·12
8 10·52 − 1·7	18 4·92 − 3
9 7·1 − 3·35	19 0·37 − 0·2
10 5·52 − 0·86	20 48 − 23·5

C You can use these methods to solve simple problems, like the ones in this exercise. Remember to check each answer using one of the methods described in Unit 10.

1 A piece of ribbon is 7 metres long, and a piece of length 1·5 metres is cut off from this. How much ribbon is left?

2 Three ball bearings weigh 6·2 grams, 5·8 grams and 7·1 grams. What is their total weight?

3 Take 9 from 14·82.

4 Increase 12·3 by 2·6.

5 A scientist weighs an empty beaker, and finds that it weighs 51·6 grams. She then pours some water into the beaker, and it now weighs 92·8 grams. What is the weight of the water in the beaker?

6 What number is 1 more than 3·8?

7 A car's petrol tank has a total capacity of 40 litres. If there are 36·4 litres of petrol in the tank, how much space is left?

8 In an ice-skating competition, the five judges awarded marks of 5·8, 5·7, 5·8, 5·8 and 5·6. What was this skater's total mark?

9 At the last census, West Germany had a population of 61·3 million and Italy had 56·7 million. How many more people had West Germany than Italy?

10 In 1954 Roger Bannister was the first man to run a mile in under four minutes; his actual time was 3 minutes 59·4 seconds. What is the difference between 59·4 seconds and 60 seconds?

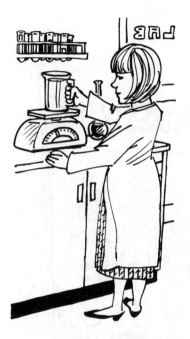

Unit 12 More Decimal Arithmetic

Multiplying decimals by whole numbers

Multiplying decimals is almost the same as multiplying whole numbers (Unit 3). Here are two ways of working out 3×7.8.

$$\begin{array}{r} 7{\cdot}8 \\ \times \quad 3 \\ \hline 23{\cdot}4 \end{array}$$

$$
\begin{aligned}
3 \times 7{\cdot}8 &= (3 \times 7) + (3 \times 8 \text{ tenths}) \\
&= 21 + 24 \text{ tenths} \\
&= 21 + 2{\cdot}4 \\
&= 23{\cdot}4
\end{aligned}
$$

A —————— Work out the following. (Remember that you can use the **commutative rule** and write the numbers the other way round if it helps.)

1 6×4.2

2 5.1×8

3 7×6.63

4 2.95×4

5 9×5.36

6 7.89×7

7 1.06×6

8 5×0.34

9 0.105×3

10 6×8.74

Multiplying decimals by decimal numbers

$3.7 \times 1.2 = ?$
a $4 \times 1 = 4$
b $37 \times 12 = 444$
c $3.7 \times 1.2 = 4.44$

$8.45 \times 0.6 = ?$
$8 \times 0.6 = 4.8$
$845 \times 6 = 5070$
so $8.45 \times 0.6 = 5.07$

This is a *little* more difficult. Suppose you want to multiply 3.7×1.2. Here are the stages:

a Round each number to 1 significant figure. Work out a rough answer; here it is 4.

b Pretend there are no decimal points and work out $37 \times 12 = 444$.

c Put back the decimal point to give a number that is 'roughly 4'. So you get 4.44.

Now look at the second example.

B

Now try these.

1 3.8×1.9	**6** 2.16×0.3
2 6.9×0.4	**7** 3.2×6.91
3 1.4×7.6	**8** 1.05×1.97
4 0.5×3.3	**9** 93.1×1.1
5 6.7×0.7	**10** 9.47×2.5

Dividing decimals by whole numbers

$4.74 \div 3 = ?$

$$3 \overline{)4.74}$$
$$1.58$$

To divide decimals, just set out the numbers as in Unit 4. Divide exactly as before, but put in the decimal point when you reach it.

C

1 $3.72 \div 6$	**6** $3.645 \div 5$
2 $5.81 \div 7$	**7** $73.808 \div 4$
3 $9.45 \div 9$	**8** $2.742 \div 3$
4 $1.84 \div 8$	**9** $0.075 \div 5$
5 $62.84 \div 4$	**10** $153.45 \div 9$

When dividing decimals, we **do not** usually write a remainder at the end. Instead, if there is something over from the last figure shown, carry it over to the next place and divide again and again, as necessary.

$$4\overline{)3 \cdot 7^{10\ 20}}$$
$$0 \cdot 9 2 5$$

Here:
$3 \div 4$ will not work. Write 0.
3 units = 30 tenths so there are 37 tenths.
$37 \div 4 = 9$ with 1 remaining.
1 tenth = 10 hundredths.
$10 \div 4 = 2$ with 2 remaining.
2 hundredths = 20 thousandths.
$20 \div 4 = 5$ exactly.

D _____

1 $52 \cdot 6 \div 4$	6 $7 \cdot 56 \div 8$
2 $3 \cdot 59 \div 2$	7 $5 \cdot 13 \div 4$
3 $7 \cdot 5 \div 3$	8 $6 \cdot 81 \div 6$
4 $3 \cdot 83 \div 5$	9 $7 \cdot 53 \div 10$
5 $2 \cdot 1 \div 4$	10 $3 \cdot 62 \div 7$

The last question shows a difficulty you sometimes meet. The remainders go on for ever without reaching the end! Usually, you stop after the **third decimal place** (with a few dots to show there is more: $0 \cdot 517 \ldots$).

Dividing by decimals

$4 \cdot 85 \div 0 \cdot 5 = ?$

a $5 \quad \div 0 \cdot 5 = 10$
b $485 \div 5 \quad = 97$
c $4 \cdot 85 \div 0 \cdot 5 = 9 \cdot 7$

Some people find dividing by decimals tricky. But you can always succeed if you use the **same method as for multiplying**:
a Approximate to 1 significant figure and get a rough answer.
b Forget the decimal point and divide.
c Replace the decimal point with the help of your rough answer.

E _____ Now try these.

1 $4 \cdot 8 \div 0 \cdot 4$	6 $1 \cdot 74 \div 0 \cdot 3$
2 $3 \cdot 5 \div 0 \cdot 2$	7 $6 \div 1 \cdot 2$
3 $5 \cdot 94 \div 0 \cdot 9$	8 $3 \cdot 06 \div 0 \cdot 1$
4 $6 \cdot 37 \div 0 \cdot 7$	9 $0 \cdot 5 \div 0 \cdot 2$
5 $1 \cdot 28 \div 0 \cdot 4$	10 $0 \cdot 36 \div 0 \cdot 4$

Using your calculator

Decimal multiplication and division is easily done on a calculator. But always remember to make a rough estimate to check your answer.

F_____ In these examples, give your answers to two decimal places:

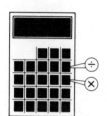

1 $31 \cdot 2 \times 4 \cdot 8$

2 $4 \cdot 832 \div 7$

3 $14 \cdot 4 \div 16$

4 $7 \cdot 28 \times 0 \cdot 08$

5 $3 \cdot 56 \div 1 \cdot 33$

6 $4 \cdot 446 \times 25$

7 $182 \div 12$

8 $16 \cdot 2 + 4 \cdot 17$

9 $11 \cdot 53 \times 9 \cdot 52$

10 $7 \cdot 2 \div 11 \cdot 52$

11 $7 \times 5 \cdot 81$

12 $6 \cdot 7 \times 1 \cdot 35$

13 $3 \cdot 61 \div 1 \cdot 45$

14 $57 \cdot 3 \times 2 \cdot 96$

15 $48 \cdot 24 - 0 \cdot 6$

16 $3 \cdot 57 \div 2 \cdot 1$

17 $2 \cdot 841 \times 7 \cdot 5$

18 $9 \cdot 91 \times 3 \cdot 48$

19 $(2 \cdot 73)^2$

20 $5 \cdot 35 \div 6 \cdot 49$

Did your rough estimate help you to avoid mistakes?

G_____ This exercise is of a different type.

Choose any two-digit number: 48, say.

Multiply its digits together: $4 \times 8 = 32$.

If the answer has two digits, multiply **them** together: $3 \times 2 = 6$.

Go on doing this until you get a single-digit answer, then stop.

So starting with 48 we get the chain
$$48 \rightarrow 32 \rightarrow 6$$
which has three numbers altogether.

Different starting numbers will give you longer or shorter chains.

1 Can you find a starting number that gives a chain of **four** numbers?

2 There is one two-digit number (and only one) that gives a chain of five numbers – can you discover which number it is?

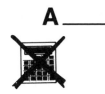

Unit 13 | Problems with Decimals

It is important to be able not only to add, subtract, multiply and divide decimals but also to decide **when** to do each of these.

A ──────

For each of these examples, write down what **operation** you would use (+, −, × or ÷). Then go back and do as many of the examples as you have time for. Remember to do a rough check for each answer. Do **not** use a calculator for this exercise.

1   A plank is 3·84 metres long. If it is cut into six shorter pieces (each the same length) how long will each piece be?

2   A small farm has three fields, whose areas are 2·6 acres, 1·7 acres, and 0·95 acres. What is their total area?

3   If £1 is worth 1·4 dollars, what is the value in dollars of £5?

4   A 1-litre measuring jug contains 0·65 litres of milk. How many litres of water could be added to make the total volume up to exactly 1 litre?

5   In the school sports, Myfanwy cleared 1·33 metres in the high jump. If the old record was 1·29 metres, by how much did she beat the record?

6   The heights of three pupils were respectively 1·66 metres, 1·69 metres, and 1·75 metres; what was their average height?

7   What would be the total weight of 100 nails each weighing 1·5 grams?

8   A roll of adhesive tape is 10 metres long. After 3·7 metres have been used, how many metres are left?

9   A metal bar of length 1·365 metres is heated, and becomes 0·013 metres longer. What is its new length?

10   If one pound (weight) is 0·454 kilograms, what would be the equivalent in kilograms of three pounds?

B ———— You may use a calculator for these questions if you wish.

1 If a 'unit' of electricity costs 5·17 pence, what would be the cost of 830 units?

2 A garage advertises petrol at 42·3 pence per litre; how much will 35 litres cost?

3 How many litres at 42·3 pence per litre could you buy for £10 (1000 pence)?

4 In June, July and August one year the rainfall at a particular place was 0·75 inches, 1·1 inches and 1·43 inches. What was the total rainfall for the three months?

5 An ordinary wine bottle holds 0·7 litres of wine. How many bottles would be needed to contain 4·9 litres?

6 If Graeme is 1·62 metres tall and Lesley is 1·7 metres tall, how much taller is Lesley than Graeme?

7 A hiker walks 6·4 miles in the morning and 8·3 miles in the afternoon; how far has she walked altogether?

8 A bag of sugar weighs 2·2 pounds; how much would ten bags weigh?

9 How many bags of sugar would you need if you wanted 14 pounds of sugar for a jam recipe?

10 Five boxes of matches contain 48, 49, 46, 48 and 47 matches. What is the average number of matches in these boxes?

The last question in exercise **B** above shows that an average need not always be a whole number; there was an average of 47·6 matches in these boxes, even though no box actually contained that number of matches. In the same way, we are told that the average size of a family in Britain is 2·3 children, though of course there are no families anywhere which have decimal children! The next exercise gives more illustrations of this.

C ———— 1 A cricketer scores 37, 22, 59, 0, 47 and 0 runs in his first six games of a season; what is his average score?

2 A girl takes several maths tests and her scores (out of 10) are 9, 9, 8, 10, 9, 10, 10, 7, 10 and 10; what is her average score?

3 A survey of cars passing along a road showed the following numbers of occupants: 1, 3, 1, 1, 4, 1, 2, 1, 1, 1, 2, 2, 1, 5, 1, 3, 2, 1, 4. What was the average number of occupants?

4 A class register shows the following attendances for last week: 22, 24, 23, 23, 19. What was the average daily attendance?

5 Four tins of beans are weighed carefully, and their contents weigh 448 grams, 451 grams, 447 grams and 449 grams. What is the average weight of the contents of each tin?

Unit 14 | Vulgar Fractions

A **vulgar fraction** is part of a whole number.

This pizza has been cut into 8 equal pieces. Each piece is one eighth ($\frac{1}{8}$) of the pizza.

2 pieces of pizza are $\frac{2}{8}$

3 pieces of pizza are $\frac{3}{8}$

Each fraction has two parts, a number on the bottom and a number on the top.

The number on the bottom is the **denominator**. It tells us how many parts the pizza was cut into.

Examples of other denominators are: thirds (3), fifths (5), tenths (10), fifteenths (15), and so on.

The number on the top is the **numerator**. It tells us how many pieces of pizza we have got.

Two fractions are so common that they have special names. $\frac{1}{2}$ is called one **half** and $\frac{1}{4}$ is called one **quarter**.

A _____ Write these fractions in figures.

1 two fifths
2 one seventh
3 three tenths
4 two thirds
5 one half

6 three quarters
7 two ninths
8 five eighths
9 one twentieth
10 fifteen sixteenths

B _____ Write these fractions in words.

1 $\frac{1}{3}$ 3 $\frac{1}{6}$ 5 $\frac{1}{4}$ 7 $\frac{11}{20}$ 9 $\frac{1}{12}$

2 $\frac{4}{5}$ 4 $\frac{2}{7}$ 6 $\frac{7}{8}$ 8 $\frac{7}{16}$ 10 $\frac{15}{32}$

C _____ Write down (in figures) the fraction of each of these shapes that has been shaded. For example, the answer to Question 1 is ½.

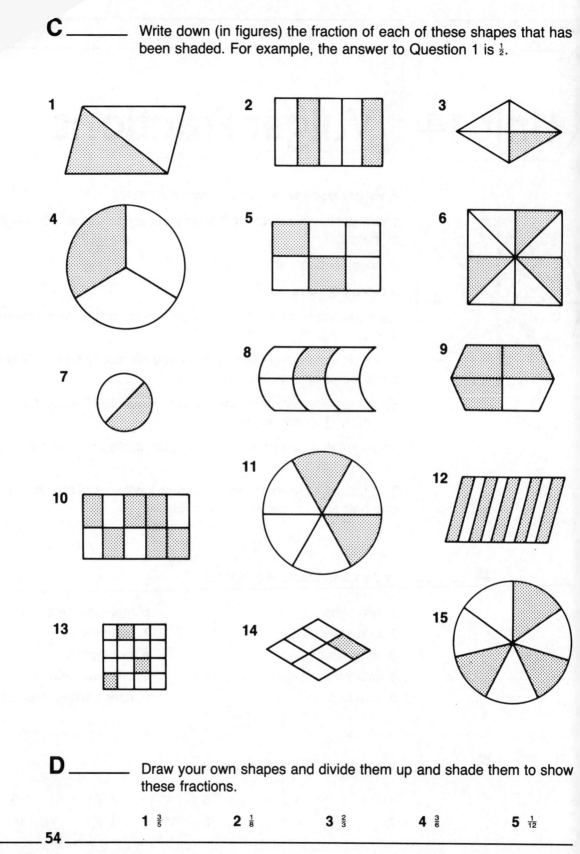

D _____ Draw your own shapes and divide them up and shade them to show these fractions.

1 $\frac{3}{5}$ **2** $\frac{1}{8}$ **3** $\frac{2}{3}$ **4** $\frac{3}{6}$ **5** $\frac{1}{12}$

Equivalent fractions

Look at these three diagrams of pies. The shaded areas are $\frac{1}{2}$, $\frac{2}{4}$ and $\frac{3}{6}$. They are all the same size, so $\frac{1}{2} = \frac{2}{4} = \frac{3}{6}$.

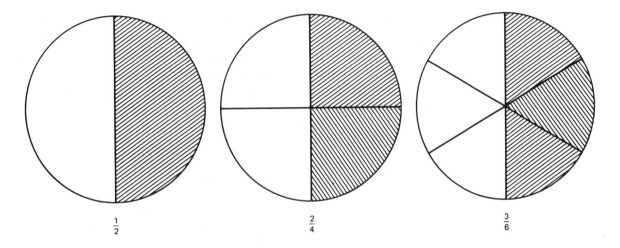

$$\frac{1}{2} \qquad \frac{2}{4} \qquad \frac{3}{6}$$

Suppose you divided three more pies of the same size into 8, 10 and 12 parts. $\frac{4}{8}$, $\frac{5}{10}$ and $\frac{6}{12}$ are all half the pie.

It is often very useful to write one fraction in another way, as an **equivalent fraction**.

Suppose you want to write 3 quarters in eighths. If the pie is divided into 8 pieces, there are twice as many pieces as there are when it is divided into quarters. But each is only half the size. So, to get the same amount, you need twice as many pieces, so $\frac{3}{4} = \frac{6}{8}$.

Suppose you want to write $\frac{1}{3}$ in twelfths. If the pie is divided into twelve pieces instead of three, we have four times as many pieces ($3 \times 4 = 12$). To cut the same amount of pie you need 4 pieces instead of one, so $\frac{1}{3} = \frac{4}{12}$.

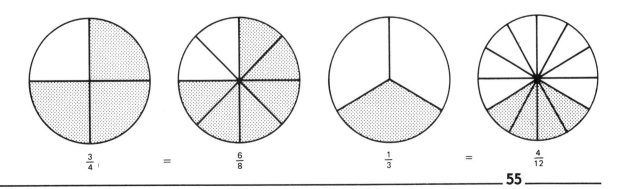

$$\frac{3}{4} \quad = \quad \frac{6}{8} \qquad\qquad \frac{1}{3} \quad = \quad \frac{4}{12}$$

E ———

1 Write $\frac{1}{2}$ in sixths
2 Write $\frac{2}{3}$ in sixths
3 Write $\frac{1}{4}$ in eighths
4 Write $\frac{1}{2}$ in tenths
5 Write $\frac{2}{3}$ in twelfths

6 Write $\frac{3}{4}$ in eighths
7 Write $\frac{2}{5}$ in tenths
8 Write $\frac{1}{6}$ in twelfths
9 Write $\frac{3}{8}$ in sixteenths
10 Write $\frac{1}{4}$ in twentieths

In the same way, we can change fractions in the other direction, taking **fewer** pieces rather than **more**. Suppose we want to change $\frac{6}{10}$ into fifths. We are going to divide the pie into five pieces rather than ten — that is, there will be only half as many pieces – so we shall need to take only half as many (three instead of six) to get the same amount of pie. So $\frac{6}{10} = \frac{3}{5}$.

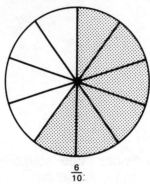
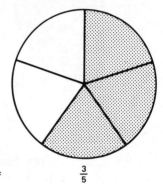

$\frac{6}{10}$ = $\frac{3}{5}$

F ———

1 Write $\frac{8}{12}$ in thirds
2 Write $\frac{3}{6}$ in halves
3 Write $\frac{2}{8}$ in quarters
4 Write $\frac{5}{15}$ in thirds
5 Write $\frac{10}{12}$ in sixths

6 Write $\frac{4}{8}$ in quarters
7 Write $\frac{12}{20}$ in tenths
8 Write $\frac{12}{20}$ in fifths
9 Write $\frac{14}{16}$ in eighths
10 Write $\frac{18}{24}$ in quarters

Unit 15 | Equivalent Fractions

$$\frac{5}{10} = \frac{1}{2}$$
$$\frac{16}{24} = \frac{8}{12} = \frac{2}{3}$$
$$\frac{20}{24} = \frac{10}{12} = \frac{5}{6}$$

We have already seen how one fraction can be written as an equivalent fraction.

Usually we try to write fractions in their **lowest terms**. This means making the denominator and numerator as small as possible.

Reducing a fraction to its smallest terms is called **simplifying** (or **cancelling**) the fraction.

$$20 \div 4 = 5$$
$$24 \div 4 = 6$$
$$\text{So } \frac{20}{24} = \frac{5}{6}$$

Suppose you want to reduce $\frac{20}{24}$ to its lowest terms. Find the **highest common factor** for 20 and 24. This is 4. Now divide both 20 and 24 by 4. This gives $\frac{20}{24} = \frac{5}{6}$, the lowest terms.

A _____ Reduce these fractions to their lowest terms.

1 $\frac{6}{8}$	3 $\frac{12}{15}$	5 $\frac{14}{21}$	7 $\frac{30}{50}$	9 $\frac{25}{60}$
2 $\frac{6}{9}$	4 $\frac{20}{25}$	6 $\frac{24}{36}$	8 $\frac{16}{48}$	10 $\frac{21}{56}$

Improper fractions

If a fraction is bigger than 1 whole it can be written in two ways.

$$1\tfrac{1}{2} = \frac{2}{2} + \frac{1}{2} = \frac{3}{2}$$

$1\frac{1}{2}$ is a **mixed** number.

$\frac{3}{2}$ is an **improper** fraction.

Quite often, it is useful to change a mixed number to an improper fraction or an improper fraction to a mixed number.
Here $\frac{7}{3}$ is being changed to a mixed number, using the idea that
3 thirds is 1 whole, so
6 thirds is 2 wholes, and 1 third left.

$$\frac{7}{3} = \frac{6+1}{3} = 2\tfrac{1}{3}$$

You may find it simpler to say $\frac{7}{3}$ is $7 \div 3$.
$7 \div 3 = 2$ remainder 1. The remainder is $\frac{1}{3}$ left as a fraction.
Changing the other way is easy. Here there are 5 wholes and each whole has 4 quarters. 5 wholes $= 5 \times 4 = 20$ quarters so there are $20 + 1 = 21$ quarters ($\frac{21}{4}$).

$$\frac{7}{3} = 7 \div 3 = 2 \text{ rem } 1$$
$$= 2\tfrac{1}{3}$$

$$5\tfrac{1}{4} = \frac{20+1}{4} = \frac{21}{4}$$

B _____ Write these improper fractions as mixed numbers.

1 $\frac{9}{5}$ **3** $\frac{10}{3}$ **5** $\frac{21}{10}$ **7** $\frac{24}{11}$ **9** $\frac{27}{10}$

2 $\frac{11}{4}$ **4** $\frac{12}{5}$ **6** $\frac{4}{3}$ **8** $\frac{31}{12}$ **10** $\frac{35}{8}$

C _____ Write these mixed numbers as improper fractions.

1 $2\frac{1}{4}$ **3** $5\frac{1}{2}$ **5** $1\frac{3}{4}$ **7** $7\frac{2}{3}$ **9** $1\frac{1}{6}$

2 $4\frac{2}{3}$ **4** $2\frac{3}{5}$ **6** $3\frac{3}{8}$ **8** $6\frac{1}{4}$ **10** $2\frac{2}{3}$

Changing decimals to vulgar fractions

In Unit 9 you learnt that the first column after the decimal point gives tenths.

The second column gives hundredths, and so on.

Here, some decimals have been written as fractions.

In **c** the fractions have been cancelled to their lowest terms.

a $0.1 = \frac{1}{10}$ $0.2 = \frac{2}{10}$ $0.3 = \frac{3}{10}$

b $0.01 = \frac{1}{100}$ $0.02 = \frac{2}{100}$ $0.03 = \frac{3}{100}$

c $0.35 = \frac{35}{100} = \frac{7}{20}$ $0.24 = \frac{24}{100} = \frac{6}{25}$

D _____ Write these decimals as vulgar fractions in their lowest terms:

1 0·9   **3** 0·4   **5** 0·5   **7** 0·34   **9** 0·75

2 0·7   **4** 0·8   **6** 0·63   **8** 0·15   **10** 0·125

Changing vulgar fractions to decimals

$$\frac{3}{8} = 8\overline{)3.000}$$
$$0.375$$

The easiest way to convert a vulgar fraction into a decimal is by simple division. $\frac{3}{8}$ is the same as $3 \div 8$ and you get your answer as shown. (Try it on a calculator to check.)

E _____ Write these vulgar fractions as decimals. Stop after three decimal places if necessary.

1 $\frac{2}{5}$ **3** $\frac{3}{4}$ **5** $\frac{1}{4}$ **7** $\frac{5}{8}$ **9** $\frac{1}{6}$

2 $\frac{1}{2}$ **4** $\frac{7}{10}$ **6** $\frac{3}{10}$ **8** $\frac{2}{3}$ **10** $\frac{5}{9}$

Recurring decimals

$\frac{2}{3} = 0.6666666666666$
$= 0.\dot{6}$

The answers to the last three questions in Exercise **E** show how even simple fractions can give answers that go on for ever. These are called **recurring** decimals, because the same figures keep repeating.

There is a special way of showing these. Simply **put a dot over the figure that repeats**.

If two figures repeat like this: 0·272727 put dots over them both 0·2̇7̇.

F _____ **Learn by heart** the decimal equivalents of $\frac{1}{4}$, $\frac{1}{2}$ and $\frac{3}{4}$; you will need these over and over again.

G _____ Use a calculator to work out the decimal equivalents of $\frac{1}{7}$, $\frac{2}{7}$, $\frac{3}{7}$ and $\frac{4}{7}$; write them all down in full, leaving off the last figure on the calculator. Can you guess what the answers to $\frac{5}{7}$ and $\frac{6}{7}$ will be before you check them on the calculator?

H _____ Use a calculator to work out the decimal equivalents of all the elevenths, from $\frac{1}{11}$ up to $\frac{10}{11}$. Write them down. Look at your answers – can you see an easy way of remembering all these? (*Hint:* it will help if you know your multiplication tables!)

Addition of Fractions

Adding fractions

$\frac{2}{5}+\frac{1}{5}=\frac{3}{5}$

This problem is easy. 2 fifths plus 1 fifth is 3 fifths, just like 2 bananas plus 1 banana is 3 bananas.

$\frac{2}{3}+\frac{1}{6}=\frac{5}{6}$

Adding $\frac{2}{3}+\frac{1}{6}$ is more difficult. These fractions are of different kinds. You couldn't add 2 goldfish to 1 cat and get a *sensible* answer. *Not* 3 goldfish, *not* 3 cats. Not three of anything but maybe 1 well fed cat! So you need to think more carefully about this.

The **only** way to add fractions is to make each fraction into the same kind. That is, **make the fractions have the same denominator**.

$\frac{2}{3}=\frac{4}{6}$, so you can rewrite the sum as $\frac{4}{6}+\frac{1}{6}$.

$\frac{2}{3}+\frac{1}{6}=\frac{4}{6}+\frac{1}{6}=\frac{5}{6}$

$\frac{5}{8}+\frac{3}{4}=?$

$\frac{5}{8}+\frac{6}{8}=\frac{11}{8}=1\frac{3}{8}$

$\frac{4}{5}+\frac{1}{4}=?$

$\frac{16}{20}+\frac{5}{20}=\frac{21}{20}=1\frac{1}{20}$

$\frac{5}{8}+\frac{3}{4}$ is a similar example, easily solved when the denominator is 8 for both.

This example is more difficult. Both fractions must be changed to get the same denominator. The lowest common multiple (Unit 7) of 4 and 5 is 20.

$\frac{4}{5}=\frac{16}{20}$ and $\frac{1}{4}=\frac{5}{20}$ and you get the answer as shown.

A

Work out the following, writing answers in their lowest terms.

1 $\frac{1}{5}+\frac{3}{5}$ 5 $\frac{1}{2}+\frac{1}{4}$ 9 $\frac{5}{8}+\frac{3}{4}$ 13 $\frac{1}{2}+\frac{1}{3}$ 17 $\frac{2}{3}+\frac{5}{6}$

2 $\frac{3}{8}+\frac{1}{8}$ 6 $\frac{3}{16}+\frac{7}{16}$ 10 $\frac{2}{9}+\frac{4}{9}$ 14 $\frac{3}{4}+\frac{1}{2}$ 18 $\frac{1}{2}+\frac{2}{3}$

3 $\frac{1}{4}+\frac{3}{4}$ 7 $\frac{1}{2}+\frac{3}{8}$ 11 $\frac{4}{5}+\frac{1}{10}$ 15 $\frac{7}{10}+\frac{1}{4}$ 19 $\frac{5}{16}+\frac{1}{2}$

4 $\frac{1}{4}+\frac{5}{8}$ 8 $\frac{3}{5}+\frac{1}{10}$ 12 $\frac{1}{4}+\frac{2}{3}$ 16 $\frac{7}{8}+\frac{1}{4}$ 20 $\frac{3}{5}+\frac{5}{8}$

Subtracting fractions

$\frac{7}{8} - \frac{3}{8} = \frac{4}{8} = \frac{1}{2}$

$\frac{5}{8} - \frac{1}{2} = \frac{5}{8} - \frac{4}{8} = \frac{1}{8}$

These simple examples show the method of subtraction, which is almost the same as addition.

B ———— Work out the following.

1. $\frac{9}{10} - \frac{7}{10}$ 3. $\frac{3}{4} - \frac{1}{16}$ 5. $\frac{2}{3} - \frac{4}{9}$ 7. $\frac{1}{2} - \frac{2}{5}$ 9. $\frac{1}{3} - \frac{1}{5}$

2. $\frac{5}{6} - \frac{2}{3}$ 4. $\frac{3}{5} - \frac{1}{5}$ 6. $\frac{7}{8} - \frac{1}{2}$ 8. $\frac{3}{4} - \frac{1}{8}$ 10. $\frac{3}{4} - \frac{1}{3}$

Mixed numbers

The easiest way of dealing with mixed numbers is to deal separately with the whole number parts, then the fractional parts, as shown here.

$$2\tfrac{1}{2} + 4\tfrac{3}{4} = (2+4) + \tfrac{1}{2} + \tfrac{3}{4}$$
$$= 6 + 1\tfrac{1}{4}$$
$$= 7\tfrac{1}{4}$$

$$5\tfrac{2}{3} - 2\tfrac{1}{4} = (5-2) + \left(\tfrac{2}{3} - \tfrac{1}{4}\right)$$
$$= 3 + \tfrac{5}{12}$$
$$= 3\tfrac{5}{12}$$

In the example below we have a slight difficulty because $\left(\frac{4}{8} - \frac{7}{8}\right)$ cannot be done without using negative numbers. This is easily overcome by changing one of the three wholes to $\frac{8}{8}$.
$(3 = 2 + 1 = 2 + \frac{8}{8})$.

$$4\tfrac{1}{2} - 1\tfrac{7}{8} = (4-1) + \left(\tfrac{1}{2} - \tfrac{7}{8}\right)$$
$$= 3 + \left(\tfrac{4}{8} - \tfrac{7}{8}\right)$$
$$= 2 + \tfrac{8}{8} + \tfrac{4}{8} - \tfrac{7}{8}$$
$$= 2\tfrac{5}{8}$$

C ———— Work out the following.

1. $1\tfrac{1}{2} + 3\tfrac{1}{4}$ 3. $4\tfrac{1}{2} - 1\tfrac{1}{4}$ 5. $3\tfrac{3}{4} - 2\tfrac{1}{8}$ 7. $3\tfrac{1}{4} - 1\tfrac{1}{2}$ 9. $8 - 3\tfrac{3}{8}$

2. $2\tfrac{1}{3} + 1\tfrac{1}{3}$ 4. $2\tfrac{2}{3} + 4$ 6. $5\tfrac{1}{2} + 1\tfrac{1}{2}$ 8. $6\tfrac{2}{5} - 2\tfrac{3}{4}$ 10. $3\tfrac{2}{3} + 4\tfrac{1}{5}$

D _____

1 If $\frac{3}{4}$ of a loaf of bread has been eaten, what fraction is left?

2 From a 3 metre length of tape, $2\frac{1}{2}$ metres have been cut away; what length remains?

3 How many cupfuls (each $\frac{1}{3}$ of a pint) will be needed to fill a 2 pint container?

4 Bernard is taking part in a marathon race. After he has run $\frac{3}{5}$ of the distance, what fraction has he still to run?

5 Three books are in a pile. One is $2\frac{1}{2}$ inches thick, one is $3\frac{1}{4}$ inches, and the other is 3 inches. How thick is the pile?

6 What is $7\frac{3}{5}$ to the nearest whole number?

7 Maria gets 15 marks out of 20 in a test. Write this mark as a fraction in its lowest terms.

8 If three-eighths of a class are boys, what fraction of the class are girls?

9 Two pieces of wood have to be fastened together. One piece is $\frac{3}{4}$ inch thick, and the other is $\frac{9}{16}$ inch thick. How thick are they together?

10 How many quarter-pound packets of tea would weigh the same as a seven-pound carton?

$\frac{1}{4}$ lb packets of tea

Unit 17

Multiplying Fractions

Here is a way you might use to show a small child how to multiply 3×4.

The oblong is 4 squares wide and 3 squares long. There are 12 squares altogether so $4 \times 3 = 12$.

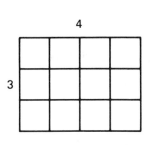

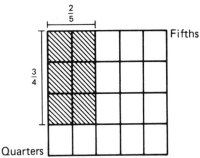

Quarters

Fifths

$$\frac{2}{5} \times \frac{3}{4} = ?$$

You can look at multiplication of fractions in the same way. Suppose you want to multiply $\frac{2}{5} \times \frac{3}{4}$.

This square is 1 unit by 1 unit. The length is marked off in 5 parts (fifths). The side is marked off in 4 parts (quarters).

Look at the shaded area where $\frac{2}{5}$ and $\frac{3}{4}$ coincide. It consists of 6 small parts out of twenty in the whole square.
So $\frac{2}{5} \times \frac{3}{4} = \frac{6}{20}$

Top line $2 \times 3 = 6$
Bottom line $5 \times 4 = 20$
$\frac{2}{5} \times \frac{3}{4} = \frac{6}{20} = \frac{3}{10}$

Looking at the numbers involved, you see that we can get the answer without drawing a square. Simply **multiply the numerators and then the denominators** like this. Of course, you write the answer in its lowest terms.

A ———— Use **diagrams** to work these out.

1 $\frac{1}{2} \times \frac{3}{4}$ 2 $\frac{5}{6} \times \frac{1}{3}$ 3 $\frac{1}{5} \times \frac{1}{2}$ 4 $\frac{2}{3} \times \frac{1}{4}$ 5 $\frac{3}{8} \times \frac{4}{5}$

B ———— Work these out by direct multiplication.

1 $\frac{3}{8} \times \frac{1}{2}$ 3 $\frac{1}{10} \times \frac{3}{4}$ 5 $\frac{6}{7} \times \frac{7}{8}$ 7 $\frac{2}{3} \times \frac{3}{5}$ 9 $\frac{9}{10} \times \frac{2}{3}$

2 $\frac{1}{3} \times \frac{5}{9}$ 4 $\frac{1}{2} \times \frac{1}{7}$ 6 $\frac{4}{5} \times \frac{3}{8}$ 8 $\frac{5}{8} \times \frac{4}{9}$ 10 $\frac{11}{12} \times \frac{3}{4}$

Multiplying mixed numbers

Method 1

$2\frac{1}{2} \times 3\frac{4}{5} =$ $(2 \times 3\frac{4}{5}) + (\frac{1}{2} \times 3\frac{4}{5})$

$= (2 \times 3) + (2 \times \frac{4}{5}) + (\frac{1}{2} \times 3) + (\frac{1}{2} \times \frac{4}{5})$

$= 6 \ + \ \frac{8}{5} \ + \ 1\frac{1}{2} \ + \ \frac{4}{10}$

$= 6 \ + \ \frac{16}{10} \ + \ 1\frac{5}{10} \ + \ \frac{4}{10}$

$= 7 \ + \ \frac{25}{10}$

$= 7 \ + \ 2\frac{5}{10}$

$= 9\frac{1}{2}$

Method 2

$2\frac{1}{2} \times 3\frac{4}{5} = \frac{5}{2} \times \frac{19}{5}$

$= \frac{95}{10}$

$= 9\frac{5}{10}$

$= 9\frac{1}{2}$

When mixed numbers are involved, you have the choice of the two methods shown here.

Method 1 uses the **distributive rule** (Unit 3).

In method 2 we change mixed numbers into improper fractions (Unit 15).

You should use the one that you find easiest. Usually, Method 1 is best if doing an easy problem in your head. Method 2 is best when writing problems down.

C _____ Use either method to work out these. (Write 4 as $\frac{4}{1}$ if you are using Method 2.)

1 $1\frac{1}{2} \times \frac{2}{5}$ **3** $\frac{3}{5} \times 4\frac{1}{2}$ **5** $1\frac{3}{5} \times 2\frac{1}{8}$ **7** $1\frac{2}{3} \times 6$ **9** $5 \times 3\frac{2}{3}$

2 $2\frac{1}{4} \times \frac{3}{4}$ **4** $1\frac{1}{4} \times 3\frac{1}{3}$ **6** $3\frac{1}{2} \times 4$ **8** $1\frac{1}{2} \times 1\frac{3}{4}$ **10** $6\frac{1}{4} \times \frac{9}{10}$

Fractions of whole numbers

$\frac{3}{4}$ of 24?

$\frac{3}{4} \times \frac{24}{1} = \frac{72}{4} = 18$

or $\frac{1}{4}$ of 24 $= 6$

$\frac{3}{4}$ of 24 $= 3 \times 6$

$= 18$

With fractions, 'of' means 'times' so you can write $\frac{3}{4} \times \frac{24}{1}$ and go on as before.

Or you can find one quarter (1/4) of 24 and then multiply by 3.

The second method is easiest for simple numbers.

D _____ Work out the following by either method.

1 $\frac{1}{8}$ of 32

2 $\frac{2}{3}$ of 15

3 $\frac{1}{5}$ of 40

4 $\frac{5}{8}$ of 48

5 $\frac{1}{4}$ of 16

6 $\frac{3}{4}$ of 72

7 $\frac{4}{5}$ of 35

8 $\frac{5}{6}$ of 42

9 $\frac{3}{5}$ of 110

10 $\frac{7}{10}$ of 50

E _____ Get about ten strips of paper, each about 3 cm wide and 30 cm long. (The exact measurements are not important.) Take the first strip, and _without using a ruler_ fold it exactly into halves – this is very easy.

Take the second strip and fold it into quarters; fold the third strip into eighths – these are easy too.

Take the next strip and try to fold it into thirds – much harder; if you can manage that, fold the next strip into sixths.

Use the remaining strips to try fifths, sevenths, ninths, tenths and even elevenths if you have any paper left. Which do you find the most difficult?

F _____

1 A factory employs 160 people. If three quarters of these are women, how many men are employed at the factory?

2 If there are 635 Members of Parliament, and 245 of these are Labour Party members, is this more or less than half?

3 A survey shows that three fifths of all cats like 'Pussidin' cat food. If a hundred cats are tested, how many of them should like 'Pussidin'?

4 $\frac{3}{4}$ of the children in a class said they had been abroad for their holidays, and half of those said they had been to France. What fraction _of the whole class_ had been to France?

5 When a Roman legion was 'decimated', it meant that one tenth of all the soldiers were executed. If a particular legion started with 600 soldiers, how many were left after it had been decimated?

Unit 18 | Dividing Fractions

Division is the inverse (opposite) of multiplication.

We look at division of fractions in this way:

How many thirds make 4?

There are 3 thirds in 1, so 4 contains 12 thirds.

So $4 \div \frac{1}{3} = 12$

$4 \div \frac{1}{3} = 12$

$4 \times 3 = 12$

But $4 \times 3 = 12$ so it seems that dividing by $\frac{1}{3}$ is the same as multiplying b 3.

Here is a second example: $\frac{3}{4} \div \frac{1}{8}$.

$\frac{3}{4} \div \frac{1}{8} = ?$

$\frac{6}{8} \div \frac{1}{8} = 6$

$\frac{3}{4} \times 8 = 6$

$\frac{3}{4} = \frac{6}{8}$ so $\frac{3}{4} \div \frac{1}{8} = 6$.

But $\frac{3}{4} \times 8$ is the same as $\frac{3}{4}$ of 8 which is also 6, so dividing by $\frac{1}{8}$ is the sam as multiplying by 8.

This third example is more difficult.

$\frac{7}{8} \div \frac{3}{4} = ?$

$\frac{7}{8} \div \frac{6}{8} = \frac{7}{6}$

$\frac{7}{8} \times \frac{4}{3} = \frac{28}{24} = \frac{7}{6}$

$\frac{7}{8} \div \frac{3}{4}$ means how many lots of $\frac{3}{4}$ make $\frac{7}{8}$?

$\frac{3}{4} = \frac{6}{8}$ so we can write the sum $\frac{7}{8} \div \frac{6}{8}$.

Of course 7 eighths divided by 6 eighths is $\frac{7}{6}$. But to get the same answe faster just multiply by $\frac{4}{3}$, which is called the **reciprocal** of $\frac{3}{4}$.

A simple rule

All these examples show that we can change any division by a fractio into an equivalent multiplication.

The rule is: **dividing by a fraction is equivalent to multiplying by its reciprocal** (what you get by turning it upside down).

These examples show how the rule is applied.

$$4 \div \frac{1}{2} = 4 \times \frac{2}{1} = 4 \times 2 = 8$$
$$15 \div \frac{1}{3} = 15 \times \frac{3}{1} = 15 \times 3 = 45$$
$$15 \div \frac{3}{4} = 15 \times \frac{4}{3} = \frac{60}{3} = 20$$
$$\frac{5}{8} \div \frac{1}{2} = \frac{5}{8} \times \frac{2}{1} = \frac{10}{8} = \frac{5}{4} = 1\frac{1}{4}$$
$$\frac{1}{5} \div \frac{3}{8} = \frac{1}{5} \times \frac{8}{3} = \frac{8}{15}$$

1 $\frac{3}{4} \div \frac{1}{2}$	**6** $\frac{3}{4} \div \frac{6}{7}$
2 $\frac{5}{6} \div \frac{2}{3}$	**7** $\frac{3}{5} \div \frac{4}{5}$
3 $\frac{5}{9} \div \frac{3}{4}$	**8** $\frac{7}{12} \div \frac{5}{8}$
4 $\frac{3}{5} \div \frac{7}{8}$	**9** $\frac{1}{6} \div \frac{2}{3}$
5 $\frac{3}{4} \div \frac{3}{8}$	**10** $\frac{5}{9} \div \frac{7}{3}$

Problems with mixed numbers

The easiest way to deal with mixed numbers is to change them into improper fractions first, like this:

$$4\tfrac{1}{2} \div 5\tfrac{2}{5} = \tfrac{9}{2} \div \tfrac{27}{5} = \tfrac{9}{2} \times \tfrac{5}{27} = \tfrac{45}{54} = \tfrac{5}{6}$$

B ——————— Now try these.

1 $3\tfrac{1}{2} \div 2\tfrac{1}{4}$	**3** $5\tfrac{2}{3} \div 2\tfrac{1}{8}$	**5** $2\tfrac{2}{3} \div 4$	**7** $2\tfrac{1}{4} \div \tfrac{3}{4}$	**9** $7 \div 1\tfrac{1}{2}$
2 $1\tfrac{1}{3} \div 1\tfrac{3}{5}$	**4** $1\tfrac{3}{4} \div \tfrac{7}{8}$	**6** $6\tfrac{7}{8} \div 3\tfrac{1}{4}$	**8** $\tfrac{4}{5} \div 3\tfrac{1}{3}$	**10** $\tfrac{11}{16} \div 3$

C ———————

1 How many pieces of ribbon, each $4\tfrac{1}{2}$ inches long, could be cut from a 36 inch length?

2 What is $\tfrac{1}{4}$ of 3? What is $3 \div \tfrac{1}{4}$? Are they the same? They shouldn't be.

3 At a certain school each lesson lasts $\tfrac{3}{4}$ hour. How many lessons are there in a 'working day' of $5\tfrac{1}{4}$ hours?

4 Some wooden planks are $1\tfrac{1}{2}$ cm thick; how many such planks would be needed to give a total thickness of at least 10 cm?

5 If $\tfrac{3}{4}$ pound of sweets is shared equally between six children, how much (as a fraction of a pound) will each of them get?

Which fraction is larger?

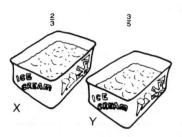

Here are two equal-sized containers of ice cream, **X** and **Y**. **X** is $\frac{2}{3}$ full, is $\frac{3}{5}$ full. Which contains the most?

The way to find out is to convert both to the same kind of fraction (se Unit 16). Converting both to fifteenths we get:

$\frac{2}{3} = \frac{10}{15}$ and $\frac{3}{5} = \frac{9}{15}$

X therefore contains $\frac{1}{15}$ more than **Y**.

D _____ Find the larger of the two fractions in each case:

1 $\frac{3}{4} : \frac{5}{8}$ **3** $\frac{7}{12} : \frac{5}{8}$ **5** $\frac{1}{2} : \frac{8}{15}$ **7** $\frac{7}{8} : \frac{5}{7}$ **9** $\frac{7}{9} : \frac{4}{5}$

2 $\frac{1}{3} : \frac{3}{10}$ **4** $\frac{3}{8} : \frac{1}{3}$ **6** $\frac{2}{7} : \frac{1}{3}$ **8** $\frac{2}{5} : \frac{1}{4}$ **10** $\frac{5}{7} : \frac{7}{10}$

Comparing fractions and decimals

$\frac{1}{3}$ and $0 \cdot 4$

$0 \cdot 3$ and $0 \cdot 4$

or $\frac{1}{3}$ and $\frac{4}{10}$, $\frac{1}{3}$ and $\frac{2}{5}$, $\frac{5}{15}$ and $\frac{6}{15}$

To compare fractions with decimals, you can either convert vulga fractions to decimal form (see exercise 15E) or convert decimals int fractions (see exercise 15D).

In the next exercise, do whichever you think is easier.

E _____ Find the larger in each case.

1 $\frac{1}{4} : 0 \cdot 2$ **3** $\frac{1}{3} : 0 \cdot 3$ **5** $\frac{2}{3} : 0 \cdot 7$ **7** $\frac{2}{5} : 0 \cdot 36$ **9** $\frac{1}{2} : 0 \cdot 4$

2 $\frac{7}{10} : 0 \cdot 8$ **4** $\frac{3}{8} : 0 \cdot 5$ **6** $\frac{5}{8} : 0 \cdot 65$ **8** $\frac{7}{9} : 0 \cdot 80$ **10** $\frac{3}{4} : 0 \cdot 7$

F _____

An old cowboy died and left his string of 11 horses to his three children The oldest child (said the cowboy's will) was to get half the horses, the second child one quarter of them, and the third child one sixth. Just as the three heirs were wondering how to divide up the horses withou killing any of them, the sheriff rode up and showed them the way to do it. What was the sheriff's solution?

G _____ Without writing anything down, work out

$\frac{1}{2} \times \frac{2}{3} \times \frac{3}{4} \times \frac{4}{5} \times \frac{5}{6} \times \frac{6}{7} \times \frac{7}{8} \times \frac{8}{9} \times 9$

Unit 19 Percentages

In everyday life, a very common form of fraction is the **percentage**. This is simply a fraction whose denominator is 100 but which is written slightly differently.

Sometimes percentages contain fractions and decimals.

17 per cent (or 17%) is $\frac{17}{100}$
25 per cent is $\frac{25}{100}$ or $\frac{1}{4}$

$$37\frac{1}{2}\% = \frac{37\frac{1}{2}}{100} = \frac{75}{200} = \frac{3}{8}$$

$$4\cdot5\% = \frac{4\cdot5}{100} = \frac{45}{1000} = \frac{9}{200}$$

A _____ Write the following as vulgar fractions in their lowest terms.

1 30%	**3** 80%	**5** 12%	**7** 31%	**9** $18\frac{3}{4}\%$
2 55%	**4** 10%	**6** 28%	**8** $12\frac{1}{2}\%$	**10** $6\cdot4\%$

Changing vulgar fractions to percentages

Because a percentage is 'out of 100', 100% is one whole. We use this fact when changing fractions to percentages.

1 whole = 100%
$\frac{3}{4} = \frac{3}{4}$ of 100% = 75%
$\frac{3}{8} = \frac{3}{8}$ of 100% = $\frac{300}{8}$ %
   = 37·5 %

B _____ Write the following as percentages:

1 $\frac{1}{2}$	**3** $\frac{7}{10}$	**5** $\frac{4}{5}$	**7** $\frac{2}{3}$	**9** $\frac{3}{7}$
2 $\frac{3}{5}$	**4** $\frac{1}{4}$	**6** $\frac{5}{8}$	**8** $\frac{3}{16}$	**10** $\frac{1}{6}$

Changing decimal fractions to percentages

$45\% = \frac{45}{100} = 0.45$
$20\% = \frac{20}{100} = 0.20$
$0.62 = \frac{62}{100} = 62\%$
$0.3 = \frac{30}{100} = 30\%$

In decimal fractions, the second decimal place represents hundredths. So conversion is very easy, as these examples show.

C _____ Write these percentages as decimals.

1 36%	3 23%	5 72%	7 5%	9 $22\frac{1}{2}\%$
2 41%	4 90%	6 22%	8 50%	10 $3\frac{1}{2}\%$

D _____ Write these decimals as percentages:

1 0.35	3 0.21	5 0.02	7 0.04	9 0.325
2 0.47	4 0.82	6 0.8	8 0.1	10 0.075

E _____ By writing the fraction or decimal as a percentage, say which is the larger of each of these pairs.

1 $40\% : \frac{1}{2}$ 6 $72\% : 0.7$

2 $35\% : 0.04$ 7 $28\% : \frac{1}{4}$

3 $87\% : \frac{8}{9}$ 8 $95\% : 0.99$

4 $12\% : 0.2$ 9 $17\% : \frac{1}{6}$

5 $6\% : \frac{1}{16}$ 10 $7\% : 0.05$

Some uses of percentages

It is often useful to be able to work out (say) what 25 per cent of 80 really is. Suppose you are earning £80 a week and your boss offers you the choice of a 25 per cent rise or an extra £25 a week – which should you choose?

25% of $80 = \frac{25}{100} \times \frac{80}{1}$ (because 'per cent' means 'out of 100'; and 'of' with
a fraction is equivalent to 'times')

$$= \frac{2000}{100}$$
$$= 20$$

The same method will work for all calculations of this kind; a pocket calculator is very useful, but you should still write down the steps shown above.

F_____ Work out these percentages.

1 30% of 500
2 25% of 60
3 50% of 36
4 26% of 800
5 15% of 120

6 40% of 50
7 22% of 150
8 65% of 40
9 72% of 225
10 $12\frac{1}{2}$% of 56

G_____ 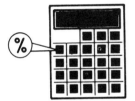 If your calculator has a button marked '%', find out (by asking your teacher or by reading the instruction booklet) how it works, and then do Exercise 19F again using that button.

H_____

1 In an opinion poll, 46% of people said 'yes' to a particular question and 35% said 'no'; what percentage said 'don't know' to that question?

2 If Jason scored 37 out of 50 in his French exam, what percentage is that?

3 In Loamshire, 60% of all the money raised by the County Council is spent on education. If Loamshire raised 500 million pounds altogether last year, how much of this went to education?

4 The rate of unemployment in Newtown is 17 per cent. If there are 5000 people of working age in Newtown, how many of them are unemployed?

5 What percentage of the people of Newtown *do* have a job?

6 In a holiday survey, 85 per cent of people said they thought Blackpool was the best place for a holiday. (The survey was done on Blackpool beach!) If 60 people were asked altogether, how many of them said that Blackpool was best?

7 In the same survey, 6 people out of the 60 said they though Scarborough was better; what percentage was this?

8 If four out of every five people can't tell Heron margarine from butter, what percentage **can** tell the difference?

9 A part-time shop worker earning £25 per week is offered a choice between a 20% pay rise and an extra £7 a week. Which should he choose?

10 Another worker in the same shop earns £40 per week and is offered the same choice (20% or £7). Which offer is better for her?

Unit 20 | Money

Most of the arithmetic done outside school or college is concerned with money in one way or another.

Either now, or when you leave school or college you will be involved in some of the money activities shown here. Throughout your life, rarely a day will go by when you do not have to do some simple, but important, arithmetic with money.

The British money system

The basic unit of money is one pound sterling (£1). This is divided into 100 pence (p). Because the system is based on 100, money arithmetic is almost exactly the same as arithmetic with ordinary numbers.

> Four pounds twenty-seven pence = £4.27
> Forty-eight pence = 48p or £0·48
> Three pounds and five pence = £3·05
> Three pounds and fifty pence = £3·50

When writing down sums of money we use the £ symbol **or** the p symbol as these examples show. We *never* use both at once.

Except for small amounts like 3p or 7p, there should always be two figures for the pence. Thus there is no confusion between amounts like three pounds and five pence and three pounds and fifty pence. If a calculator gives 3·5 as the answer to a money sum, it usually means £3·50.

A ———— Write properly in figures:

1 two pounds and sixteen pence
2 five pounds and thirty pence
3 six pounds and four pence
4 seventy-two pence
5 eight pounds and fifty pence
6 seven pounds
7 one pound and three pence
8 forty pence
9 eight pounds and a penny
10 ten pounds and twenty-five pence

Using coins

Although you can usually get change if you do not have the exact amount of money you need, there are times when the exact amount is needed. Many vending machines do not give change, for example, and some city buses insist on exact fares only.

B ———— If you could use any coins you wanted, say how you would make up exactly each of the following amounts.

1 35p   2 16p   3 63p   4 47p   5 88p

C ———— If you could not use more than one coin of each value, say how you would make up exactly each of the following amounts.

1 36p   2 58p   3 17p   4 86p   5 7p

D _____ Say how you would make up each of these amounts using as few coins as possible:

1 51p   **2** 70p   **3** 25p   **4** 42p   **5** 39p

E _____ If the only coins you had were three 10p pieces, three 5p pieces and four 2p pieces, say how you would make up each of these amounts:

1 40p   **2** 26p   **3** 13p   **4** 18p   **5** 51p

F _____ What is the largest amount (totalling more than £1) you could have in current British coins, but still *not* be able to make up exactly £1?

G _____ Lisa's mother finally got tired of the mess in her bedroom, and they came to an agreement. Every Saturday Lisa's mother would inspect the room; if it was tidy she would put 50p in a special money box, but if it was untidy she would take 20p out of the box. They kept this up for a year, but Lisa really didn't try very hard and after 52 Saturdays there was just 10p in the box. How many times had her bedroom passed inspection?

Unit 21 | Money Arithmetic

Arithmetic with money is almost the same as arithmetic with whole numbers and decimals, but there are a few small differences that need care.

Much money arithmetic is done mentally – while you are shopping for example – as in the first exercise.

A ——— Work out the following *in your head*.

1 Add 37p and 45p.
2 Take 83p from £1.
3 What is twice 35p?
4 Share £5 equally between four people.
5 Add 45p to £1.35.
6 If you buy a packet of biscuits for 63p, and pay with a £1 coin, how much change should you get?
7 What is the total price of five cans of lemonade at 18p each?
8 What would it cost to buy a packet of crisps at 12p, a chocolate biscuit at 9p and a drink at 15p?
9 If the full adult fare for a bus journey is 56p, how much is the half fare for children?
10 If you buy meat costing £2.40, and pay with a £5 note, how much change should you get?

Writing down addition and subtraction

£4·30 + 25p = ?

```
  4·30
+   25
──────
  4·55
```

£4.30 + £25 = ?

```
   4·30
+ 25·00
───────
  29·30
```

£7 − £2·35 = ?

```
  7·00
- 2·35
──────
  4·65
```

When adding money the arithmetic is easy, as long as you put the pounds and pence in their correct columns.

With a calculator, don't worry about writing numbers in columns. For £4.30 + 25p enter 4·30 + 0·25 (**not** 25 which is £25!).

Whole pounds are no trouble with a calculator, so 7 − 2·35 gives the right answer.

If the calculator gives 2·7 (meaning £2·70) or 4·5 (£4·50) you must write your answer properly – with two pence figures.

The same rules apply to subtraction. Just as in Unit 11, it is useful to fill in any missing 0's in the column, as this example shows.

Occasionally (usually when dividing), you get an answer like 3·165, which means £3·16 with a remainder. We'll return to this later.

B ———— Use *either* pencil and paper *or* a calculator to work out the following:

1 £1·44 + £2·31
2 £2·85 + £3
3 £7·63 − £2·41
4 78p + £1·45
5 £2·54 − £1·79

6 £6·36 − £2·76
7 £5·32 + £1·48
8 £5 − £3·84
9 £4·25 − 70p
10 £3·22 + 78p

Multiplying and dividing money

£2·74 × 5 = ?

```
   2·74
×     5
───────
 13·70
```

Money can only be multiplied by numbers. Money **cannot** be multiplied by money. For example £2·26 × £4 has no meaning!

Suppose you pay out £2·74 each to five people. The amount needed is £2·74 × 5 and you set it out as shown. 5 is neither pounds or pence so it need not be in either column. Just put it in a convenient place.

A calculator will show 13·7, so you will need to write it out correctly.

$£25 \div 4 = ?$

$$4 \overline{)25 \cdot 00}$$
$$6 \cdot 25$$

$£9 \cdot 40 \div 7 = ?$

$$7 \overline{)9 \cdot 40}$$
$$1 \cdot 3 \, 4 \text{ rem } 2$$

To divide sums of money, we use the same method as for decimals in Unit 12. There is one difference. With decimals, you keep dividing. With money, you cannot divide after the second column of pence. If there is anything left, round off to the nearest penny, and write a remainder.

Doing this division on a calculator would give 1·3428571, which you would round off to the nearest penny (2 decimal places), and write £1·34.

C _____ Use *either* pencil and paper *or* a calculator to work out the following.

1 £4·17 × 3   6 £3·36 ÷ 4
2 £1·35 × 8   7 £2·30 × 6
3 £3·14 ÷ 2   8 £6 ÷ 5
4 £5·40 ÷ 6   9 37p × 7
5 16p × 4   10 £1·25 × 8

D _____ Work out the following *to the nearest penny.*

1 £5·27 ÷ 4   4 £3·86 ÷ 5
2 £2·41 ÷ 3   5 £7·45 ÷ 2
3 £1·59 ÷ 7

E _____ Work out the following *showing any remainder.*

1 £2·49 ÷ 2   4 £3·78 ÷ 6
2 £13·25 ÷ 4   5 £2·48 ÷ 9
3 £7 ÷ 3

F _____

Three people went into an ironmonger's shop, each wanting to buy the same thing. Arthur bought 5, which cost him 50p; Bimla bought 20, which cost her £1; and Ravi bought 360, which cost him just £1.50 even though there was no special discount for large numbers.

What do you think they were buying? (Where they lived might have something to do with it!)

G Use the price list below to help you work out the total cost of each of the following orders.

```
— LONDON TRANSPORT CAFE MENU —      PRICES

Bacon ————————————————————————— 60 p
Eggs ————————————————————— each 40 p
Sausages ————————————————— each 60 p
Chips ————————————————————————— 50 p
Fish —————————————————————————— 80 p
Baked Beans —————————————————— 50 p
Toast ————————————————————— slice 20 p
Bread Roll ——————————————————— 15 p
Tea (small) —————————————————— 30 p
    (large) —————————————————— 40 p
Coffee (small) ——————————————— 40 p
    (large ————————————————————— 50 p
Milk ——————————————————————————— 40 p
```

1 Bacon, egg and chips, and a large tea.
2 Two sausages, chips and beans, and a glass of milk.
3 Fish, chips and beans, a bread roll and a large coffee.
4 Three rounds of toast and a small coffee.
5 Two eggs on a slice of toast, and a large tea.
6 Baked beans on toast, bacon and two eggs, and two small teas.
7 Bacon, egg and chips twice, and two small coffees.
8 Four bread rolls, three rounds of toast, and six large coffees.
9 Sausage, bacon, chips and beans, and a large tea, all three times.
10 Two eggs, chips and beans, a bread roll and a milk, all four times.

H Clare, Donna and Jasminder went for a picnic together. Clare took four sandwiches and Donna took five; Jasminder had no time to make any sandwiches, but offered the other two 90p (all she had with her) if they would share theirs with her. Clare and Donna agreed, and the three girls shared the sandwiches equally between them.

What would be the fairest way for Clare and Donna to split the 90p, bearing in mind that they took different numbers of sandwiches? Are you *sure* that your way is the fairest?

Wages and Salaries

PAY ADVICE	
Date	10/4/87
Employee Name	R Brown
N.I. Number	NB 371037 A
Tax Code	242L O
Earnings Basic	105.00
Overtime	
Bonus	
Deductions Tax	15.66
N.I. (Not Out)	9.49
Total Deductions	25.15
N.I. To Date	9.49
Gross Taxable	
Pay To Date	105.00
Tax Deducted To Date	15.66
Amount Payable	79.85

Money earned by working is usually one of the following:

A **wage** calculated at so much per hour and paid weekly.

A **salary** calculated at so much per year and paid monthly.

A **fee** is an amount paid for a particular job, charged by a self-employed person.

A **commission** is an amount paid, usually to sales staff, according to how much they sell.

Most full-time workers are not paid the amount they earn. This is because of certain **deductions**, the most common being *Income tax* and *National Insurance contributions*, taken off by the employer and paid to the Government. Sometimes there are *Pension contributions* that go into a fund to provide money for when the person retires.

We will look at some of these deductions in a later unit. Here, we'll consider only wages and salaries before deductions – the **gross** amounts earned.

A

Work out the gross weekly wage of each of the following workers.

1 Andrew works for 40 hours at £3·00 per hour.
2 Barbara works for 35 hours at £2·00 per hour.
3 Colin works for 36 hours at £2·50 per hour.
4 Dawn works for 38 hours at £2·60 per hour.
5 Eleanor works for 34 hours at £2·10 per hour.
6 Frank works for 40 hours at £1·70 per hour.
7 Gary works for 32 hours at £2·20 per hour.
8 Heather works for 36 hours at £1·75 per hour.
9 Indira works for 30 hours at £1·40 per hour.
10 Janet works for 34 hours at £1·85 per hour.

Basic week and overtime

In most firms there is an agreement about a **basic week**, for which a worker (employee) is paid a **basic wage**. A typist for example might receive £120 for a 40-hour week. Thus the **basic rate** is £3 per hour ($\frac{120}{40}$).

Hours worked above the basic week are called **overtime**. Quite often, the employee is paid at a higher rate for overtime. On a basic rate of £3 per hour,

'Time and a quarter' would be £3 × $1\frac{1}{4}$ = £3·75 per hour

'Time and a half'   would be £3 × $1\frac{1}{2}$ = £4·50 per hour

'Double time'   would be £3 × 2 = £6·00 per hour

B _____ Assuming a firm pays 'time and a half', work out the overtime rate for the following basic rates.

1 £2·00 per hour
2 £3·00 per hour
3 £2·40 per hour
4 £1·80 per hour
5 £2·60 per hour

6 £1·90 per hour
7 £2·50 per hour
8 £1·40 per hour
9 £2·10 per hour
10 £3·20 per hour

Calculating the gross wage

Once we know the overtime rate, we can work out how much is due to any worker who works more than the basic week.

Suppose that Yasmin has a basic rate of £2·30 per hour for a basic 35-hour week, but that in one particular week she works 39 hours.

She has worked 35 hours basic, plus 4 hours overtime at time and a half, so she should get 35 × £2·30 + 4 × £3·45, which is £80·50 + £13·80 or £94·30 altogether.

C _____ Work out the gross weekly wage for each of the following workers. The figures in brackets are the basic rate of pay and the length of the basic week; use the overtime rates you worked out in Exercise **B**.

1 Karen works 40 hours (£2.00 p.h., 36 hours).
2 Lawrence works 38 hours (£3.00 p.h., 35 hours).
3 Moses works 41 hours (£2.40 p.h., 36 hours).
4 Nigel works 39 hours (£1.80 p.h., 37 hours).
5 Olwen works 44 hours (£2.60 p.h., 35 hours).
6 Paul works 36 hours (£1.90 p.h., 36 hours).
7 Quentin works 40 hours (£2.50 p.h., 38 hours).
8 Rachel works 42 hours (£1.40 p.h., 40 hours).
9 Soraya works 37 hours (£2.10 p.h., 36 hours).
10 Teresa works 45 hours (£3.20 p.h., 35 hours).

Salaries

Salaries are usually quite easy to deal with, because most salaries are at a fixed annual rate. To convert this annual salary into a monthly payment we simply divide by twelve (to the nearest penny, if necessary).

D _____ Work out the gross monthly salary of each of the following:

1 Fiona earns £9300 per annum (that is, per year).
2 Graeme earns £8760 per annum.
3 Hilary earns £12 144 per annum.
4 Isabel earns £11 748 per annum.
5 John earns £7320 per annum.
6 Khalid earns £9075 per annum.
7 Lorna earns £10 266 per annum.
8 Martin earns £6495 per annum.
9 Naomi earns £8255 per annum.
10 Omar earns £9350 per annum.

More Money Problems

30% of £25?

$$\frac{30}{100} \times \frac{25}{1} = \frac{750}{100} = £7\cdot50$$

In Unit 19 (Exercise **F**) you learnt how to calculate the percentage of a number. The method is used here to find 30% of £25.

With money, there is an easier method.

$1\% = \frac{1}{100}$, so 1% of £1 = 1p. In this example 1% of £25 is 25p so 30% of £25 is $30 \times 25p = 750p = £7\cdot50$.

In the following examples, use whichever method you prefer.

A

1 25% of £200
2 30% of £500
3 10% of £150
4 15% of £30
5 20% of £45

6 5% of £10
7 12% of £55
8 40% of £18
9 7% of £1500
10 $6\frac{1}{2}$% of £40

Increasing and decreasing by a percentage

An electrician comes to your house and does the rewiring. Here is his bill. Notice that the whole charge has been increased by 15% because of value added tax (VAT). This can come as a shock!

```
Labour at £5
40 hours at £5 per hour... £200
Materials ................ £100

              Sub-total £300
            VAT at 15% £ 45

                  Total £345
```

In life you have to deal with many situations where you need to know the effect of a percentage charge. Some of these are shown here.

Here we see the simplest way to increase or decrease a number by percentage.

In the first example, find 15% of £12 and add it to £12.

In the second example, find 10% of £12 and subtract it from £12.

a £12 plus 15%

15% of £12 = $\frac{15}{100} \times 12$ = £1.80

Total = £12 + £1.80

= £13.80

b £12 less 10%

10% of £12 = $\frac{10}{100} \times 12$ = £1.20

Total = £12 − £1.20

= £10.80

B ———— Try these examples.

1 Increase £30 by 10%

2 Increase £150 by 20%

3 Increase £50 by 75%

4 Increase £85 by 8%

5 Increase £28 by 16%

6 Decrease £200 by 20%

7 Reduce £60 by 35%

8 Decrease £180 by 5%

9 Decrease £70 by 15%

10 Reduce £40 by 50%

C ———— Use the methods of the previous exercise to help you answer the following questions. Use a calculator if you wish.

1 A bill comes to £140, to which 15% value added tax must be added. What is the total amount including tax?

2 A coat is marked with a price of £55, but in a sale all prices are to be reduced by 20%. What is the sale price of the coat?

3 Damien has a job which pays £65 per week. When he is given an 8% wage rise, what is his new weekly wage?

4   Hayley has a meal in a restaurant. The meal costs £5.80, but a 10% service charge has to be added to this. How much is the service charge?

5   The Smith family have booked a foreign holiday with Snailtours, at a cost of £540. Shortly before the holiday, they are told that there will be a 5% surcharge added to the price of their holiday – what will the new price be?

6   Chris works as a shop assistant. He buys goods from the shop which are priced at £4.50, but he is entitled to a staff discount of 10%. How much does he have to pay?

7   A car loses 30% of its value in the first year it is on the road – this is called **depreciation**. A new car sells at £3450 – what is it worth after a year?

8   Lucien cashes a cheque for £50, but the hotel where he cashes it takes $1\frac{1}{2}$% commission. How much does Lucien actually get?

9   Jayne deposits £120 in a building society account. At the end of the year, the building society adds $8\frac{1}{2}$% interest to the account. How much is in the account now?

10 A second-hand car dealer buys a car for £320. If he hopes to make 20% profit on the car, what price should he sell it for?

Percentages, profit and loss

A shopkeeper or trader buys at one price and sells at a higher price. The difference is the **profit**, money needed to pay for the expenses of running the business, as well as providing enough to live on.

For example, suppose an antique dealer buys a sideboard for £600 and sells it for £780.

The profit is £780 − £600 = £180.

The percentage profit, £180 out of £600 is $\frac{180}{600} \times 100 = 30\%$

D

1. Martin buys a second-hand leather jacket for £12 and sells it f £15. How much profit has he made?

2. What is Martin's profit as a percentage of the cost price?

3. Joy buys pot plants at £2 each and sells them at a profit of 50% What is the price at which she sells them?

4. Vicki buys a second-hand record player for £8, and later sells for £6.40. How much has she lost?

5. What is her loss as a percentage of the price she paid?

6. The price of a school dinner goes up from 60p to 70p. What is th as a percentage increase, to the nearest whole number?

7. A grocer buys 20 packets of biscuits for £4.50, and sells them 27p each. How much profit does he make altogether?

8. What is this profit as a percentage of his cost price?

9. A furniture store plans to make a 60% profit on the goods it sell If the store buys a double bed for £70, what price should the be be sold at?

10. A bookseller plans to get rid of last year's annuals, and is pre pared to accept a loss of 15% on each book. If the *Tiddleywink Annual* cost the bookseller £1.20, at what price should she sell it

Interest

If you deposit money in a savings account in the post office, a bank o building society, you will be paid **interest** on your savings.

If you borrow money from a bank or building society, you will pa interest.

The amount of interest paid depends on these things:

How much money invested.

How long (the time) invested.

The **rate** of interest.

In this example, you are paid interest of £2, so your total savings are now £40 + £2 = £42.

> £40 invested for 1 year at 5% interest
> = 5% of £40
> = $\frac{5}{100}$ × 40
> = £2

E ——— Work out (a) the amount of interest, and (b) the total, after one year, on each of the following investments.

1 £50 deposited in a building society account paying 6% p.a.
2 £80 deposited in a bank account paying 5% p.a.
3 £500 deposited in a special high-interest account at 9% p.a.
4 £140 deposited in a children's savings account at $3\frac{1}{2}$% p.a.
5 £1500 deposited in a building society special account at 8·35% p.a.

Simple and compound interest

If you invested £35 for 1 year at 7% you would get £2·45 in interest.

If the rate of interest stays the same, the interest earned
  in 2 years is £4·90
  in 3 years is £7·35
  and so on.

This is called **simple interest** and assumes that you only have £40 deposited all the time. This will happen if you spend the interest each year.

(If you leave the interest in the account you will earn more, because banks and building societies use a slightly different system called **compound interest**.)

F ——— Work out (a) the amount of simple interest, and (b) the total amount, on each of the following.

1 £60 deposited at 5% p.a. for two years.
2 £110 deposited at 8% p.a. for five years.
3 £2500 deposited at $6\frac{1}{2}$ p.a. for three years.
4 £400 deposited at 4% p.a. for six months (half a year).
5 £30 deposited at $5\frac{1}{4}$% p.a. for four years.

Hire purchase (HP) and credit purchase (CP)

From a legal point of view, HP and CP are different, but the money arithmetic is just the same.

If people cannot afford to pay for an item, they can often pay by instalments over a period.

Here we have a typical arrangement.

Total paid = £30 + 12 × 19
$$= £30 + 228$$
$$= £258$$

The difference of £58 between cash and credit price is a lot, but it is the interest paid for the convenience of buying on credit.

Before buying on credit you should **always** work out the extra cost and then think about two things:

a Can I wait until I have the money and pay cash? This will save the interest. If you shop around it can save more, as some sellers offer a **discount** for cash purchases.

b Can I keep up the payments? If in doubt, don't buy!

G ——— Work out the total amount payable if each of these items are bought on credit.

1 Twelve monthly payments of £8 (no deposit).
2 Twelve monthly payments of £11·50.
3 52 weekly payments of £2.
4 24 monthly payments of £13.20.
5 £27 deposit, and twelve payments of £20.
6 £80 deposit plus 24 payments of £5·50.
7 £16·50 deposit, plus £1·20 per month for two years.
8 £35 deposit, plus 26 weekly payments of £2·40.
9 £120 deposit and 52 payments of £3·80.
10 £155 deposit plus five payments of £143 each.

H _____ Work out the total HP price of each of the following, and thus work out the difference between the hire purchase price and the cash price.

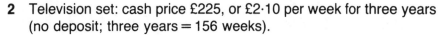

1. Bedroom suite: cash price £445, or £90 deposit and 24 monthly payments of £16·00.

2. Television set: cash price £225, or £2·10 per week for three years (no deposit; three years = 156 weeks).

3. Car: cash price £3600, or one-third deposit and 30 monthly payments of £96·50.

4. Fridge/freezer: £125 cash, or 10% deposit and £10 per month for a year.

5. Video recorder: £472 cash, or 20% down and £7·95 a week for a year (one year = 52 weeks).

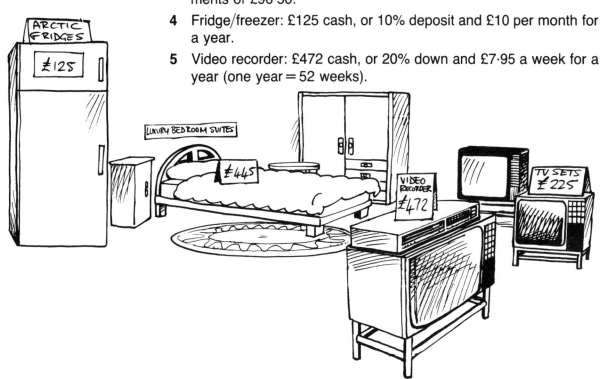

Unit 24 Look-up Tables

There are many situations in real life in which you do not work out a price by arithmetic at all – instead you look it up in a table. Such tables are designed to be easy to use, but they can still be confusing at first: this Unit will give you some simple practice.

The table below, issued by a building society, shows the monthly repayments required for a £1000 mortgage at various interest rates over various periods of time. (For amounts other than £1000, multiply the figure given by an appropriate factor. For example, you can find repayments on a £5000 mortgage by multiplying by 5.)

Years	10	15	20	25	30
Interest Rate	£	£	£	£	£
5%	10.80	8.03	6.69	5.92	5.43
6%	11.33	8.59	7.27	6.52	6.06
7%	11.87	9.15	7.87	7.16	6.72
8%	12.42	9.74	8.49	7.81	7.41
9%	12.99	10.34	9.13	8.49	8.12
10%	13.57	10.96	9.79	9.19	8.84
11%	14.16	11.59	10.47	9.90	9.59
12%	14.75	12.24	11.16	10.63	10.35
13%	15.36	12.90	11.87	11.37	11.12
14%	15.98	13.57	12.59	12.13	11.91
15%	16.61	14.26	13.32	12.90	12.70

A ——— Use the table to work out the following:

1 What is the monthly repayment on £1000 borrowed for 20 years at 8% p.a.?

2 What is the monthly repayment on £1000 borrowed for 30 years at 13% p.a.?

3 What is the monthly repayment on £1000 borrowed for 15 years at 10% p.a.?

4 *Estimate* the monthly repayment on £1000 borrowed for 20 years at 12½% p.a.

5 What is the monthly repayment on £2000 borrowed for 25 years at 6% p.a.?

6 What is the monthly repayment on £10 000 borrowed for 10 years at 11% p.a.?

7 Sam and Sheila borrow £20 000 for 25 years at 14% p.a. How much is their monthly repayment?

8 How much would Sam and Sheila repay in a year (12 months)? Use a calculator if you wish.

9 How much would they repay over 25 years?

10 If they had chosen a 30-year mortgage (instead of 25 years), how much altogether would their repayments have come to?

The table below is produced by a holiday company, and shows the price of a two-night holiday at various hotels, starting from various parts of the United Kingdom (shown as areas **A**, **B**, **C** and so on). Also shown are the extra charges for a third night at the hotel, and first-class travel on the journey. **All charges for children are half price**.

Hotel	Journey starting from area					Extra for third night
	A	B	C	D	E	
Imperial	£ 70	£ 85	£ 105	£120	£ 140	£ 25
Royal	£ 68	£ 83	£ 103	£ 118	£ 138	£ 24
Splendide Grand	£ 65	£ 80	£ 100	£ 115	£ 135	£ 22
Palace	£ 58	£ 73	£ 93	£ 108	£ 128	£ 18
De Luxe	£ 50	£ 65	£ 85	£ 100	£ 120	£ 15
Extra for first class	£ 5	£ 8	£ 10	£ 12	£ 14	

B _____ Use the table to work out the following:

1 What would be the cost of a two-night holiday at the Palace Hotel for one person travelling from area B?

2 What would be the cost of a two-night holiday at the Royal Hotel for one person travelling from area C?

3 What would be the cost of a three-night holiday at the Hotel De Luxe for one person travelling from area A?

4 What would be the cost of a two-night holiday at the Grand Hotel for one person travelling first class from area E?

5 What would be the cost of a three-night holiday at the Imperial Hotel for one person travelling first class from area D?

6 What would be the cost of a two-night holiday at the Hotel Splendide for two adults travelling from area C?

7 What would be the cost of a three-night holiday at the Royal Hotel for two adults travelling from area E?

8 What would be the cost of a two-night holiday at the Imperial Hotel for two adults and a child travelling from area B?

9 What would be the cost of a three-night holiday at the Palace Hotel for two adults and two children travelling first class from area A?

10 Suppose that you live in area B and can afford to spend up to £200 on a holiday for yourself and a friend. Which holiday will you choose? (You will have to pay the adult rates, of course.)

Some tables are not in the usual rectangular form, and take more skill to read. Probably the best known of these is the **distance table** often found in road atlases, showing the distances between various towns.

The table below shows the distances (in miles) between the main towns of the county of Cumbria.

BARROW-IN-FURNESS

74	CARLISLE									
59	26	COCKERMOUTH								
4	70	55	DALTON-IN-FURNESS							
34	47	42	30	KENDAL						
47	31	12	43	30	KESWICK					
65	28	7	61	49	19	MARYPORT				
53	21	30	49	26	18	37	PENRITH			
19	65	50	5	25	38	57	44	ULVERSTON		
51	39	13	47	55	25	14	43	46	WHITEHAVEN	
59	34	9	55	51	21	6	39	54	8	WORKINGTON

If we want to find from this table the distance between Cockermouth and Ulverston, we have to find where the Cockermouth squares cross the Ulverston squares.

The Cockermouth squares are the ones shaded diagonally; the Ulverston squares are shaded across; and you can see that the intersection is in the square numbered 50. So the distance from Cockermouth to Ulverston is 50 miles.

C_____ Use the table to find the distance between the following towns.

1   Carlisle and Kendal
2   Maryport and Workington
3   Barrow-in-Furness and Keswick
4   Dalton-in-Furness and Ulverston
5   Cockermouth and Penrith
6   Whitehaven and Carlisle
7   Ulverston and Barrow-in-Furness
8   Kendal and Whitehaven
9   Keswick and Cockermouth
10   Maryport and Penrith

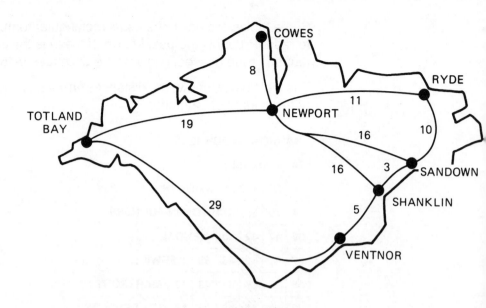

D _____ Above is a simplified map of the Isle of Wight, with distances marked in kilometres. Copy the blank distance table below, and use the map above to help you fill in the distances between the various towns. Remember always to choose the **shortest** route if you have a choice.

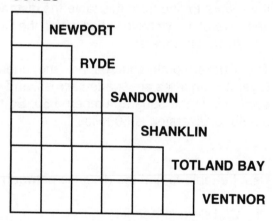

E _____ Make up a similar distance table for the main towns in your county or other local area, using a map or road signs to give you the distances.

Unit 25 | Simple Graphs

Graphs are a very important way of presenting information. You have probably seen various types in newspapers, magazines or on TV.

In this Unit and the next, we will look at some of them.

Block graphs

In the **block graph**, numbers are shown by blocks of different heights. This graph shows how many cars of each colour were found in a car park one day.

Number of cars

Colour

A ——— Use the graph above to answer the following questions.

1 How many red cars were there in the car park?
2 How many yellow cars were there?

3 Which was the most popular colour?

4 Of which colour were there five cars only?

5 There were 47 cars in the car park altogether; how many of these were none of the seven colours above, but a different colour?

B ———— With your teacher's permission, carry out a survey of the cars in your own school or college car park and draw a block graph similar to the one above. (What will you do if one of your teachers has a two-tone car?)

Pie charts

This is another kind of graph: the **pie chart**.

The circle (or 'pie') is cut into slices (**sectors**). Each sector represents different groups.

This pie chart shows how a girl divides her time each day.

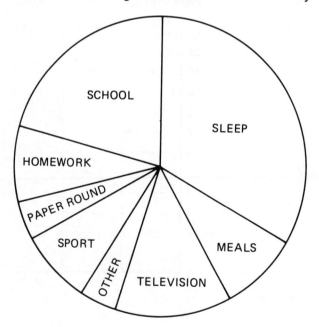

C _____ Use this pie chart to answer the following:

1 On what does the girl spend most time?
2 Does she spend more or less than a quarter of her time on school?
3 If she spends one hour a day on her paper round, how many hours does she spend on sport?
4 How many hours does she spend watching television?
5 What do you think the 'other' activities might be?

Pictograms

In a pictogram, little pictures are used to show the things being counted.

This one shows how a sports club spent its money one year. Each coin represents £10.

EQUIPMENT   £ £ £ £ £ (

TRAVELLING   £ £ £ £ £ £ £ £ £ (

PRIZES   £ £ £

REFRESHMENTS £ £ £ £ £ £ £

D _____ Look at the pictogram and answer these questions.

1 How much did the club spend on prizes?
2 How much did they spend on travelling?
3 How much did they spend altogether?
4 What percentage of the club's expenditure went on refreshments?
5 Was more or less than a quarter of the club's money spent on equipment?

More Graphs

Conversion graphs

The graph below shows the conversion between **pounds** (on the axis going across) and **dollars** (on the axis going up) and can be used to convert any amount up to ten pounds.

For example, the dotted line shows how we could use the graph to find out that £7·50 is equivalent to $10·50.

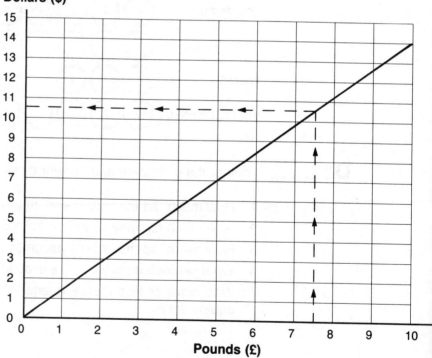

Dollars ($)

Pounds (£)

A _____ Use the graph to convert these sums.

(Work to the nearest dollar or the nearest 50p as appropriate.)

1 £5 to dollars

2 £8 to dollars

3 £2 to dollars

4 £3.50 to dollars

5 £10 to dollars

6 $14 to pounds

7 $10 to pounds

8 $1 to pounds

9 $8.50 to pounds

10 $4.50 to pounds

The second graph is similar, and shows the conversion between the Celsius (across) and Fahrenheit (up) scales of temperature.

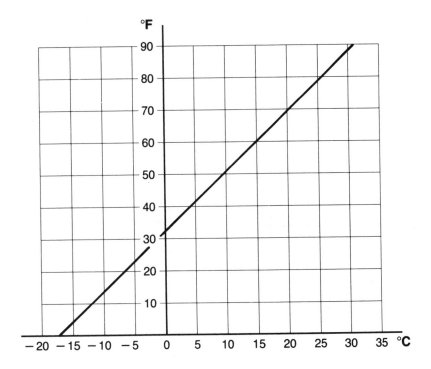

B _____ Use the graph above to convert these temperatures.

(Work to the nearest 5° as appropriate.)

1 10°C to Fahrenheit

2 −15°C to Fahrenheit

3 25°C to Fahrenheit

4 0°C to Fahrenheit

5 30°C to Fahrenheit

6 50°F to Celsius

7 80°F to Celsius

8 20°F to Celsius

9 65°F to Celsius

10 0°F to Celsius

Unit 27 Meters and Dials

It is difficult to go through life without having to read a measuring instrument.

How wide is this book? Read the rule.

How fast is the car going? Read the speedometer.

What is the patient's temperature? Read the thermometer.

How heavy is the parcel? Read the balance scales.

Is this battery OK? Read the voltmeter.

A _____ What sort of thing might you measure with each of these instruments

1 a ruler
2 a protractor
3 a stopwatch
4 a pair of scales
5 a measuring jug

6 a speedometer
7 a thermometer
8 an altimeter
9 a barometer
10 an ammeter

Reading meters

There are three ways in which meters show (display) their readings.

Digital displays
These give figures (digits) directly and are usually easy to read. The digital watch and the mileage recorder (mileometer) of a motor car are typical examples.

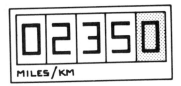

This electricity meter is another example.

Linear displays on a scale
These have an indicator moving along a straight line. Liquid thermometers and mercury barometers work like this.

Circular displays and scale
These have a pointer moving round a circular scale. Car speedometers and dial ('analogue') clocks work this way.

Linear and circular displays have one problem. Look at this speedometer. What speed does it show? You can't be sure because the point is between two marks. It is more than 35 mph but less than 40 mph. You can guess, of course, that it might be 37 mph, but this isn't very precise. All pointer and scale instruments have the problem of the pointer being between two marks.

For this reason, more and more scientific instruments are being designed to give digital read-outs.

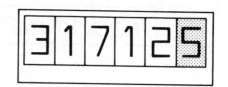

1 Look at the car mileage recorder above; the shaded figure show tenths. How far has the car travelled?

2 When we started on our present journey the reading was 31662·4 How far have we come (to the nearest mile) since then?

3 What is the current reading *to the nearest thousand miles*?

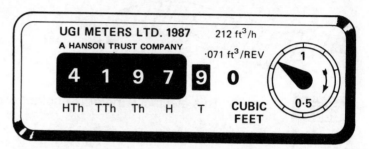

4 Look at the gas meter above, and the headings above the figures How many cubic feet of gas does it show?

5 When the meter was last read, the figures were 4178 on the let with 7 in the shaded box. How many cubic feet of gas have bee used since then?

6 If gas costs about 40p per hundred cubic feet, how much will cost for the gas used since the last reading?

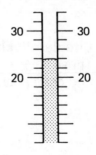

7 Look at the thermometer. How many spaces are there between 20° and 30°?

8 How many degrees does each space represent?

9 What temperature does the thermometer show at the moment?

10 Look at the measuring cylinder. How many spaces are there between 500 ml and 600 ml?

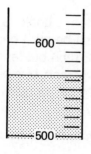

11 How many ml (millilitres) does each space represent?

12 What volume of liquid is in the measuring cylinder?

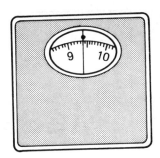

13 Look at the bathroom scales; the numbers represent stones. What is the present reading *to the nearest stone*?

14 Is the weight shown more or less than 9½ stone?

15 Given that there are 14 pounds in a stone, estimate the weight shown to the nearest pound.

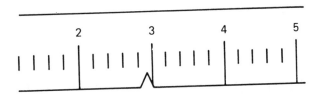

16 Look at the kitchen scales, which are numbered in kilograms. How many spaces are there between 2 kg and 3 kg?

17 How many kilograms (as a decimal) does each space represent?

18 What weight do the scales show (to one decimal place)?

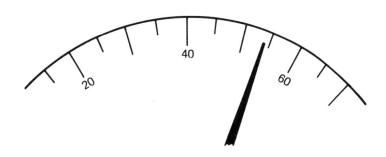

19 Look at the speedometer above, which is in miles per hour. How many mph is each space worth?

20 Estimate (to the nearest whole number) the speed in mph shown by the pointer.

Unit 28 Measuring Lengths

Units of length

METRIC   IMPERIAL

To measure lengths we use units from one of two systems.

The **imperial** system has been used in Britain for about 1000 years. It continues in use although it is gradually being replaced by the **metric** system.

The metric system was set up in France about 200 years ago. It is a decimal system and much easier to use than the imperial system. Look at the tables below and see how much easier it is to convert from one unit to another in metric measure.

Metric system (based on the metre)			
millimetre	mm		Useful conversions
centimetre	cm	10 mm = 1 cm	
metre	m	100 cm = 1 m	
kilometre	km	1000 m = 1 km	1 km = $\frac{5}{8}$ mi = 0·625 mi

Imperial system (also used in the USA)			
inch	1 in or 1″	12 inches = 1 ft	1 in = 2·54 cm
foot	1 ft or 1′	3 ft = 1 yd	1 ft = 30·48 cm = 0·30 m
yard	yd		1 yd = 0·91 m
mile	mi	1 mi = 1760 yd	1 mi = 1·6 km

A⎯⎯⎯ Which *imperial* units would you use to give the following measurements.

1 The width of this page?
2 The height of your classroom?
3 The width of a hockey pitch?
4 The distance from London to Paris?

B _____ Which *metric* units would you use to give:

1 the length of the River Severn?
2 the length of a pencil?
3 the height of these letters?
4 the length of the school hall?

5 Which sport often measures distances in *furlongs*?
6 who might measure a distance in *nautical miles*?

Using the right units

In all work with length, it is important to have a rough idea of each unit so that you can avoid silly answers such as Sean's height being 160 km instead of 160 cm.

The next exercise should help you check on your ideas; if you find it hard, or if you get many wrong answers, try to get more practice at actual measurement.

C _____ For each of the measurements below, four possible answers are given. Choose the answer that you think is most **likely** to be right.

1 The height of a fully-grown man.
   3 feet   6 feet   9 feet   12 feet

2 The length of a 50-seater bus.
   10 metres   20 metres   50 metres   100 metres

3 The thickness of a 10p piece.
   0·2 mm   2 mm   20 mm   200 mm

4 The height of an ordinary doorway.
   50 cm   100 cm   150 cm   200 cm

5 The distance from London to New York.
   5 km   50 km   500 km   5000 km

6 The length of your arm and hand.
   30 cm   30 in   30 ft   30 km

7 The height of a typical 'low bridge'.

15mm   15m   15yd   15ft

8 the distance around your waist.

60cm   60in   60mm   60ft

9 The distance an athlete could run in an hour.

12m   12mi   12cm   12mm

10 The height of an old oak tree.

25cm   25in   25ft   25km

Practising measurement

We all need to have some idea about how big things are in terms of the units used.

Use a ruler or measuring tape and do some simple measurements. Check the answers about this book given here. Measure yourself: height, length of foot, reach, etc. Then measure more objects and both small and large distances.

Object measured	Length	
	Imperial	Metric
This book: width height thickness		

Converting units

People's heights and weights are often quoted in mixed units (at least in the imperial system). For example,
height 70″ is given as 5′10″
weight 156 lb is given as 11 st 2 lb.

In general, we try to use only one unit at a time. For example, we say that this page is 189 mm wide or 18·9 cm wide, **not** 18 cm 9 mm wide.

Often you need to change from one unit to another in the same system. The next exercise will give you some practice at this.

D ———— Change these measurements

1 50 mm to centimetres
2 2 km to metres
3 $3\frac{1}{2}$ m to centimetres
4 36 mm to centimetres
5 0·45 m to millimetres

6 24 in to feet
7 4 yards to feet
8 2 miles to yards
9 $1\frac{1}{2}$ ft to inches
10 $\frac{1}{4}$ mile to yards

E ————
1 Change 5 miles into kilometres (1 mile = 1·6 km)
2 Change 400 metres into yards (1 metre = 1·1 yards)
3 Change 6 in into centimetres (1 inch = 2·54 cm)
4 Change 10 metres into inches (1 metre = 39·36 inches)
5 Change 5 ft 4 in into metres (1 inch = 2·54 cm)

Light years

Light travels at 186 000 miles per second
= 186 000 × 3600 miles per hour
= 186 000 × 3600 × 24 miles per day
= 186 000 × 3600 × 24 × 365 miles per year

For really large distances, astronomers measure in **light years**. This is the distance light travels in 1 year.

Use your calculator to check that it is about
6000 000 000 000 (6×10^{12}) miles or
9400 000 000 000 ($9\cdot4 \times 10^{12}$) kilometres.

Unit 29 Measuring Weight

Weight, like length, is measured using either the imperial or the metric system.

The table below shows all the common units.

Metric system			Conversions
gram	g	1000 g = 1 kg	
kilogram	kg	1000 kg = 1 t	1 kg = 2·2 lb
tonne	t		
Imperial system			
ounce	oz	16 oz = 1 lb	1 oz = 28 g
pound	lb	14 lb = 1 st	1 lb = 0·454 kg
stone	st	112 lb = 1 cwt	
hundredweight	cwt	20 cwt = 1 ton	
ton		1 ton = 1 tonne almost exactly	

A _____ Which **imperial** unit would you use to give the weight of the following?

1 An apple

2 A sack of potatoes

3 A chair

4 A lorry load of gravel?

Which **metric** unit would you use to give the weight of:

5 a baked bean?

6 a tin of baked beans?

7 a television set?

8 a postage stamp?

Approximately how much do _you_ weigh

9 in stones

10 in kilograms?

Estimating weights

If you want to estimate the weight of something, the easy way is to compare it with something you know the weight of **either** by picking up the known weight and the unknown weight **or** by using your memory and imagination.

B _____ Place these sports balls in order of weight, from the lightest to the heaviest: basketball, bowls, cricket, golf, table tennis, tennis, volleyball.

C _____ Look at the food in your kitchen cupboard or in a local shop. Make a list of ten different items and write the weight beside each one. For example, your list might begin:

> *packet of butter........225 grams*

D _____ Each of the items below is followed by four possible weights. Choose the weight that you think is nearest to the correct one in each case.

1 a banana

| 1 oz | 4 oz | 1 lb | 4 lb |

2 a packet of tea

| 1 g | 10 g | 100 g | 1000 g |

3 this book

| 2 oz | 12 oz | 2 lb | 12 lb |

4 an ordinary building brick

| 2 g | 20 g | 200 g | 2000 g |

5 a sheet of writing paper

| 5 g | 5 oz | 50 g | 50 oz |

If a weighing machine is available, you should use it to measure the weights of as many common objects as possible.

When you have had some practice, try *guessing* weights and using the weighing machine to check your guesses.

Just as with length, it is sometimes necessary to change from one unit of weight to another.

If we want to change (say) $3\frac{1}{2}$ lb into ounces, we simply recall that 1 lb = 16 oz, so $3\frac{1}{2}$ lb = $3\frac{1}{2} \times 16 = 56$ oz.

E _____ Change:

1 2 lb to ounces

2 3 st to pounds

3 112 lb to stones

4 24 oz to lb

5 $1\frac{1}{2}$ tons to hundredweights

6 3 kg to grams

7 5 tonnes to kilograms

8 $2\frac{1}{2}$ kg to grams

9 1500 g to kilograms

10 2300 kg to tonnes

Sometimes we need to convert units from one system to the other. Look at the conversion column on the table. We know, for example that 1 kg = 2·2 lb, so 7 kg = 7 × 2·2 = 15·4 lb.

F _____ **1** Convert 3 kg to pounds (1 kg = 2·2 lb).

2 Change 5 lb into grams (1 lb = 454 g).

3 Change $3\frac{1}{2}$ oz into grams (1 oz = 28 g).

4 Change 1 lb 4 oz into grams (1 oz = 28 g).

5 Convert 52 kg into stones and pounds (1 kg = 2·2 lb).

G _____ Young Karen was playing with some wooden blocks and a pair of scales. She found that if she put two pyramids and a sphere on one side of the scales they would just balance a cube on the other side. Then she tried a sphere and a cube together, and noticed that they balanced nicely against three pyramids. Finally she put a single pyramid on one side of the scales: how many spheres would she need on the other side to make them balance?

H _____ Try these questions. Look out for tricks!

1 Which is heavier, a pound of feathers or a pound of lead?

2 Which is worth more, 1 kg of 10p pieces or 2 kg of 5p pieces?

3 Does a pound of pure water weigh more or less when it is frozen?

4 If an orange weighs 100 g plus half an orange, how much does an orange actually weigh?

5 Estimate the weight of the earth in a hole 2 metres long, 1 metre wide and 1½ metres deep.

Unit 30 Measuring Capacity

The amount a container will hold, usually of a liquid, is called its **capacity**. The same two systems of measurement are in use.

Metric system			Some conversions
{cubic centimetre	cm³		
{or millilitre	ml	1 cm³ = 1 ml	
centilitre	cl	1 cl = 10 ml	
litre	l	1 l = 100 cl = 1000 ml	1 l = 1¾ pints
Imperial system			
fluid ounce	fl oz	20 fl oz = 1 pint	1 fl oz = 28 ml
gill		1 gill = ¼ pint	
pint		2 pints = 1 quart	
quart		4 quarts = 1 gallon	
gallon	gall		1 gall = 4½ l

A

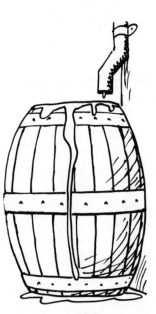

Collect a teaspoon, a dessertspoon, a tablespoon, a teacup, and a milk bottle. You will also need a graduated measuring jug or cylinder. Find a supply of water and use your equipment to answer the following questions.

1 How many teacupfuls are equivalent to a pint?
2 How many teaspoonfuls are equivalent to a tablespoonful?
3 How many dessertspoonfuls are equivalent to a teacupful?
4 How many fluid ounces or millilitres does a teacup hold?
5 Complete this table:

Container	Fl oz	ml
Teaspoon		5
Dessertspoon		
Tablespoon		
Teacup		
Milk bottle	20	

You could add some other containers to your list.

B _____
1. Change 2 pints into fluid ounces.
2. Change 12 pints into gallons.
3. Change $4\frac{1}{2}$ litres into millilitres.
4. Change 70 centilitres into litres.
5. Change 5 litres into pints (1 litre = 1·75 pints).
6. Change 40 litres into gallons (1 litre = 1·75 pints).
7. How many 70 cl wine bottles can be filled from a gallon jar (1 gallon = 4·6 litres)?
8. A recipe says 'Pour 6 fl oz milk into a jug and make it up to half a pint with cold water'. How much water has to be added?
9. Which holds more, a 1-pint milk bottle or a 70 cl wine bottle?
10. Which is the better buy: petrol at £2·04 per gallon, or petrol at 43p per litre?

C _____
The camp cook is preparing a recipe which calls for exactly four litres of water. Unfortunately, the only containers he can find are two large pans, one holding exactly three litres and the other exactly five litres.

• Using just these two pans and water from the tap, how can the cook measure out the four litres he needs?

D _____
Collect about six or eight medium-sized containers of different shapes. You might include a jam jar, an empty baked-beans tin, a milk jug and so on. By looking at them, try to arrange them in order of capacity, from the one with the smallest capacity to the one with the largest; write down the order you finally choose. Now use a measuring jug, or some other marked container, and measure the capacity of each container by filling it with water or with dry sand.

• How close was your estimated order to the correct answer?

Unit 31 Measuring Time

There is only one system of measurement for time.

The unit of time is **the second** (s). The second is based on incredibl accurate nuclear clocks. For us, just think of it as the time it takes to sa 'one little second'. For most people, when resting, the time betwee pulse beats is just under a second.

1 minute (min) = 60 s
1 hour (h)   = 60 min = 3600 s
1 day   = 24 h

The other units used are the **week**, **month** and **year** (Unit 33).

A ——— You will need a partner and a watch that measures to the neares second. Give the watch to your partner and let him or her keep tim while you try to estimate exactly one minute by silent counting or som similar method. See how close you can get; then change over and you keep time while your partner does the estimation.

B ——— Find your pulse by feeling on the inside of your wrist, just at the base of your thumb. When you have found it, count how many times it beat in one minute. The 'normal' pulse beat for most people is about 7(beats per minute, but this can easily go up to 120 or so after vigorou exercise. You could try taking your pulse again after running on th spot for a minute.

Over the centuries, people have used many different instruments fo measuring time. There have been candles which burned at a know speed, hourglasses with sand pouring from one bulb to the other sundials (you can see them in gardens or on public buildings) and othe instruments. Today we normally use a clock or a watch to measure th time, and this clock or watch may be either **digital** or **analogue**.

A digital clock is one which shows the time in figures, perhaps as 3:25 or as 18:40, while an analogue clock normally shows time by the position of its hands on a dial. Analogue clocks almost always work on a 12-hour system (where 12 o'clock is followed by 1 o'clock) but many digital clocks use a 24-hour cycle, counting right up to 23:59 before starting again at 0:00.

12-hour and 24-hour clocks

Suppose it is *a quarter to eight* – time to get up for school. This time may be shown as:

7.45 as in the *TV Times*

7:45 as on a digital clock

07 45 as on a railway timetable.

At noon the time is 12.00 noon. $7\frac{3}{4}$ hours later (*a quarter to eight*) the time may be shown as 7.45 in the *TV Times*.

19:45 as on a digital clock

19 45 as on a railway timetable (12 + 7.45 hours)

We are so used to reading dial clocks that show 12 hours and saying 6 am (before noon) and 6 pm (after noon) that many people find the 24-hour clock a problem. It is easy, as long as you remember that:

a Time starts at midnight, so after 12 noon you continue to count the hours from midnight – so 1.0 pm becomes 13.00, 2.0 pm is 14.00 and so on.

To find a 12-hour clock time from a 24-hour clock time, simply subtract 12. E.g. 16.30 − 12 = 4.30.

b Minutes are always measured after the hour, so five minutes to eight is 07.55.

c There are always two figures for the minutes, and usually two for the hours as well.

C———— What do these times mean in ordinary language?

1 08 30

2 09 15

3 07 10

4 12 05

5 14 20

6 22 00

7 15 45

8 17 25

9 00 30

10 20 35

D — How would these times appear on a railway timetable?

1 9 o'clock am (**am** stands for *ante meridiem*, which means 'before midday')
2 9 o'clock pm (**pm** stands for *post meridiem*, 'after midday')
3 half past ten in the morning
4 a quarter past seven in the evening
5 a quarter to eight in the evening
6 10 past 11 am
7 5 to 6 pm
8 9.20 am
9 3.35 pm
10 midnight

MONDAYS TO FRIDAYS

	*A	125		125✕	125✕	*A	125✕ A B	125✕ ℗ A C	*D
Derby	——	0532	——	0634	——	0717	——	0755	0802
Nottingham	——	——	0603	——	0700	——	0745	——	——
Loughborough	——	0550	0623	0652	——	——	0759	——	0821
Leicester	0327	0605	0638	0705	0727	0746	0812	——	0836
Market Harborough	——	0620	——	0720	0742	——	——	——	——
Kettering	0355	0630	0706	0731	0752	0814	——	——	——
Wellingborough	0403	0638	0715	0738	0800	0823	——	——	——
Bedford	0420	0654	——	——	——	——	——	——	——
Luton	0452b	0722b	——	——	——	——	——	——	——
London St Pancras	0537c	0735	0816	0830	0852	0923	0929	0935	——

	125✕ A	125✕ A	*D		125 I	125 I	125✕	125✕	125	125✕
Derby	——	0827	——	——	0923	——	——	1123	——	——
Nottingham	0815	——	0828	0915	——	1030	——	——	1145	1250
Loughborough	0832	——	0849	——	0942	1044	1142	——	1304	
Leicester	0845	0853	0905	0941	0955	1057	1155	1209	1317	
Market Harborough	——	0908	——	1004	——	1112	——	1224	——	
Kettering	0907	——	1004	——	1123	1217	——	1340		
Wellingborough	——	0923	——	1022	1130	1225	1239	1347		
Bedford	——	——	——	1038	——	1241	——	——		
Luton	——	0951	——	1036	1122b	1158	1322b	——	1416	
London St Pancras	1004	1019	——	1110	1125	1232	1328	1337	1450	

	125✕	125✕	125	125✕	125✕	125	*A FO	*D	125
Derby	1324	——	1423	——	——	1532	1540	——	1637
Nottingham	——	1350	——	1520	——	——	1548	——	——
Loughborough	——	——	1442	——	1534	——	——	1606	1653
Leicester	1351	1416	1455	1510	1548	1559	1610	1624	1706
Market Harborough	——	1431	——	1524	——	——	——	1640	——
Kettering	——	1442	——	1535	——	1621	——	1658	——
Wellingborough	——	1449	——	1542	——	1629	——	——	——
Bedford	——	1506	——	1559	——	——	——	——	1747
Luton	1443	1553b	——	1615	——	——	——	——	1803
London St Pancras	1517	1553	1617	1643	1705	1721	1746	——	1833

SATURDAYS

	*A	125	*	125	125	125	*D	125	125	125
Derby	——	——	0637	——	0755	0802	——	0856	——	
Nottingham	——	0615	——	0732	——	——	0843	——	0940	
Loughborough	——	0632	0656	0749	——	0821	0900	——	0957	
Leicester	0327	0645	0709	0802	0822	0836	0913	0929	1010	
Market Harborough	——	——	0724	0817	——	——	——	0944	——	
Kettering	0355	0707	0735	0827	——	——	0935	0954	1032	
Wellingborough	0403	0715	0742	0835	——	——	0943	1002	1040	
Bedford	0420	0731	——	——	0903	——	——	1018	——	
Luton	0452b	——	0810	0902	0952b	——	——	1052b	1107	
London St Pancras	0537c	0812	0838	0930	0944	——	1035	1059	1141	

	125	125	125	125	125	125	125	125	125
Derby	0956	——	1123	——	1324	——	1423	——	1532
Nottingham	——	1110	——	1250	——	1350	——	1520	——
Loughborough	——	——	1142	1304	——	——	1442	1534	——
Leicester	1022	1134	1155	1317	1351	1416	1455	1548	1559
Market Harborough	——	1149	——	——	——	1431	1510	——	——
Kettering	——	1159	——	1340	——	1441	1520	1611	——
Wellingborough	——	1207	——	1347	——	1449	1528	1618	——
Bedford	——	1223	——	——	——	1505	1544	——	——
Luton	1114	1252b	1247	1415	1443	1552b	1622b	——	1651
London St Pancras	1148	1310	1321	1449	1517	1553	1631	1710	1720

	*D	125	125	125	125	*D	125	125 *	125 *
Derby	——	1637	——	1756	——	——	1907	2018	——
Nottingham	1548	——	1830	——	1854	——	——	——	2030
Loughborough	1606	——	1726	——	1847	1913	1926	2037	2044
Leicester	1624	1703	1740	1822	1900	1929	1940	2050	2057
Market Harborough	1640	——	1755	——	1915	——	——	——	2112
Kettering	1658	1726	1803	——	1925	——	2002	——	2123
Wellingborough	——	1733	1813	——	1933	——	2010	——	2130
Bedford	——	1750	——	——	1949	——	——	——	2147
Luton	——	1822b	——	1914	2005	——	2037	——	2222b
London St Pancras	——	1830	1905	1942	2033	——	2105	——	2228

E — What time (in the 12-hour system) is shown on each of the following clocks?

1

2

3

4

5

F _____ Think of an ordinary clock with a minute hand and an hour hand. At twelve o'clock the minute hand lies directly over the hour hand; the same thing happens just after five past one, just after ten past two, and so on through the day.

• How many times in twelve hours (say, between 6 o'clock am and 6 o'clock pm) will it happen altogether? Think carefully before you answer!

G _____ • If you had two old-fashioned hourglass-type egg timers, one measuring four minutes and the other seven minutes, how could you use them (and no other instruments) to measure a time of nine minutes? (This is quite a hard problem to solve.)

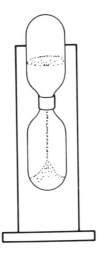 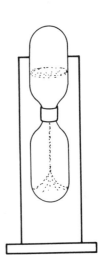

Unit 32

Times and Timetables

```
CHANNEL 5 TELEVISION — THURSDAY
4.00      News Headlines
4.05      Children's TV: Animal Fun
                        : Treasure Island
                        : Puppet Theatre
5.15      The IQ Quiz
5.45      The Joe Soap Show
6.10      Sports Report
7.00      Main News
          Weather Forecast
7.25      Music To Go Deaf By
8.00      In Your Garden
8.30      Life In Outer Mongolia
9.00      Film: Godfather Part VI
10.55     Tooth And Nail
11.40     Late News
          Evening Prayers
          Closedown
```

A ———— Look at the television schedule above, and use it to answer thes
questions:

1 At what time does *The Joe Soap Show* begin?

2 How long does it last?

3 Which programme starts at half past eight?

4 Which programme starts at five to eleven?

5 How long does the main film run for?

6 If *Animal Fun* lasts 15 minutes, and *Treasure Island* half an hour
   at what time does *Puppet Theatre* begin?

7 If the *Main News* runs for 22 minutes, how long is left for the
   Weather Forecast?

8 If the *Late News* runs for 6 minutes, and the *Evening Prayers* fo
   3 minutes, how many minutes before midnight is the *Closedown*?

9 Alicia starts watching *The IQ Quiz* at the beginning, and stays
   tuned in until the end of *Music To Go Deaf By*. How long has she
   been watching altogether?

10 Brendan watches *Life in Outer Mongolia* and the *Film*, but falls
   asleep ten minutes after the start of *Tooth and Nail*. How many
   minutes' television has he seen?

Oxford to Cambridge							
Oxford	——	07 30	08 30	10 30	13 00	15 30	18 30
Buckingham	——	08 05	——	11 05	13 35	16 05	19 05
Milton Keynes	——	08 20	——	11 20	13 50	16 20	19 20
Bedford	07 35	08 45	09 40	11 45	14 15	16 45	19 45
Cambridge	08 15	09 25	10 20	12 25	14 55	17 25	20 25

Cambridge to Oxford							
Cambridge	07 00	08 00	09 30	11 15	13 30	16 15	19 00
Bedford	07 40	——	10 10	11 55	14 10	16 55	19 40
Milton Keynes	08 05	——	10 35	——	14 35	17 20	——
Buckingham	08 20	——	10 50	——	14 50	17 35	——
Oxford	08 55	09 40	11 25	13 05	15 25	18 10	——

B ——— Use the imaginary railway timetable above to help you answer the following questions.

1 At what time does the first train leave Oxford in the morning?

2 At what time does this train reach Cambridge?

3 At what time does the last train leave Cambridge in the evening?

4 Where does this train end its journey?

5 If I catch a train from Oxford at 1 o'clock pm, at what time will I reach Bedford?

6 If I catch a train from Cambridge at quarter past four, at what time will I get to Milton Keynes?

7 If I catch the five-to-twelve train from Bedford, at what time should I get to Oxford?

8 If I am now in Bedford, and have to be in Cambridge some time before midday, which trains might I catch?

9 If I start in Milton Keynes, and have to be in Oxford by 2 o'clock, which trains might I catch?

10 If I cannot leave Oxford before four o'clock, what is the earliest that I can get to Cambridge?

11 How long does the 13 00 train from Oxford take to reach Buckingham?

12 How long does the 10 10 train from Bedford take to reach Milton Keynes?

13 How long in minutes is the journey from Bedford to Cambridge?

14 On the 08 00 express, how long is the journey from Cambridge to Oxford?

15 If I start in Oxford, travel to Cambridge by the first possible train of the day and return to Oxford by the last possible train at night, how long (in hours and minutes) have I to spend in Cambridge?

C —————— When you are cooking the Sunday dinner for the family, there are quit a number of jobs that have to be done.

* Prepare meat	2 min	* Peel potatoes	10 min
* Cut and wash cabbage	5 min	* Mix Yorkshire pudding ...	5 min
Warm up oven	10 min	Heat cabbage water	7 min
Roast meat	90 min	Roast potatoes	90 min
Boil cabbage	8 min	Bake Yorkshire pudding	30 min
* Carve meat	6 min	* Make gravy	2 min
* Serve meat	1 min	* Serve potatoes	1 min
* Serve cabbage	2 min	* Serve Yorkshire pudding	1 min
* Serve gravy	1 min		

The jobs marked * are those that keep you busy; you cannot do more than one of these at once. On the other hand, you *could* be mixing the Yorkshire pudding, say, while the meat was roasting. Some jobs mus obviously be done before others: you must peel the potatoes before you cook them, and cook them before you serve them, for example

Plan your timetable for preparing, cooking and serving this meal assuming that the family want to start eating at 1.00 pm exactly. They will, of course, expect everything to be still hot when they start to eat

Time zones

The earth rotates once every 24 hours. This means that the sun rises and the day begins at different times in different places. For example, when we are getting up in this country, it is still the middle of the night in New York.

The world is divided into time zones. The time in each zone is based on one standard time in Greenwich, England. This is known as **Greenwich Mean Time** (GMT) and is always written as a 24-hour clock time.

San Francisco	−8 hours
New York	−5 hours
Barbados	−4 hours
Paris	+1 hour
Lusaka	+2 hours
New Delhi	+5½ hours
Tokyo	+9 hours

The table shows the time ahead (+) or behind (−) GMT for various other places. Use it to answer the questions in the next exercise.

D _____ Assuming that London is on GMT (and not Summer Time),

1. When it is 5.00 pm in London, what time is it in Lusaka?
2. When it is 9.00 am in London, what time is it in Tokyo?
3. When it is 19 30 in London, what time is it in San Francisco?
4. When it is 14 15 in London, what time is it in Barbados?
5. When it is 12 45 in London, what time is it in New Delhi?
6. When it is 16 00 in London, what time is it in Paris?
7. When it is 9.15 am in Tokyo, what time is it in London?
8. When it is 11.30 pm in San Francisco, what time is it in London?
9. When it is 15 00 in Lusaka, what time is it in Barbados?
10. When it is 08 30 in Paris, what time is it in New Delhi?

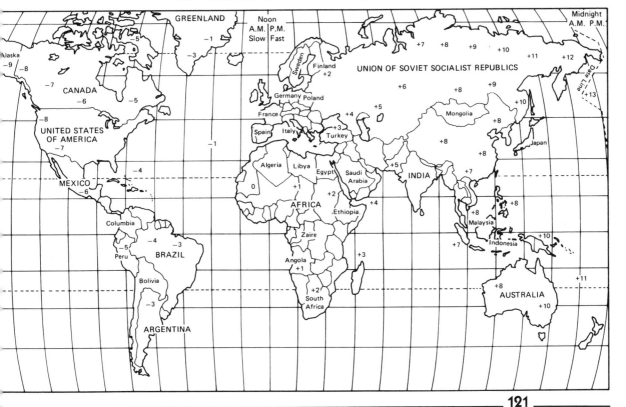

Unit 33 | The Calendar

The earth goes round the sun in about $365\frac{1}{4}$ days. In most years we have 365 days. To take in the $\frac{1}{4}$ day we have a **leap year** of 366 days every fourth year. Leap years are those that divide by 4 (like 1984, 1988, 1992, etc.)

For convenience, the year is divided into 12 **months**. They are not all the same length because 365 does not divide exactly by 12.

31 days	30 days	28/29 days
January	April	February
March	June	
May	September	
July	November	
August		
October		
December		

A **week** is seven days. 52 weeks = 52 × 7 days = 364 days. This is one or two days less than a year. You should be able to see why dates do not fall on the same day every year.

Many people use this rhyme to help them to remember how many days there are in each month. Do you know it?

30 days has September
April, June and November.
All the rest have 31 –
Except for February alone;
It has 28 days clear
29 in each leap year.

A **calendar** is the most useful way of recording dates. One for 1988 is shown on the opposite page.

CALENDAR FOR 1988

	Su	Mo	Tu	We	Th	Fr	Sa		Su	Mo	Tu	We	Th	Fr	Sa
JAN						1	2	**JUL**						1	2
	3	4	5	6	7	8	9		3	4	5	6	7	8	9
	10	11	12	13	14	15	16		10	11	12	13	14	15	16
	17	18	19	20	21	22	23		17	18	19	20	21	22	23
	24	25	26	27	28	29	30		24	25	26	27	28	29	30
	31								31						
FEB		1	2	3	4	5	6	**AUG**		1	2	3	4	5	6
	7	8	9	10	11	12	13		7	8	9	10	11	12	13
	14	15	16	17	18	19	20		14	15	16	17	18	19	20
	21	22	23	24	25	26	27		21	22	23	24	25	26	27
	28	29							28	29	30	31			
MAR		1	2	3	4	5		**SEP**					1	2	3
	6	7	8	9	10	11	12		4	5	6	7	8	9	10
	13	14	15	16	17	18	19		11	12	13	14	15	16	17
	20	21	22	23	24	25	26		18	19	20	21	22	23	24
	27	28	29	30	31				25	26	27	28	29	30	
APR						1	2	**OCT**							1
	3	4	5	6	7	8	9		2	3	4	5	6	7	8
	10	11	12	13	14	15	16		9	10	11	12	13	14	15
	17	18	19	20	21	22	23		16	17	18	19	20	21	22
	24	25	26	27	28	29	30		23	24	25	26	27	28	29
									30	31					
MAY	1	2	3	4	5	6	7	**NOV**		1	2	3	4	5	
	8	9	10	11	12	13	14		6	7	8	9	10	11	12
	15	16	17	18	19	20	21		13	14	15	16	17	18	19
	22	23	24	25	26	27	28		20	21	22	23	24	25	26
	29	30	31						27	28	29	30			
JUN				1	2	3	4	**DEC**					1	2	3
	5	6	7	8	9	10	11		4	5	6	7	8	9	10
	12	13	14	15	16	17	18		11	12	13	14	15	16	17
	19	20	21	22	23	24	25		18	19	20	21	22	23	24
	26	27	28	29	30				25	26	27	28	29	30	31

A _____ Use the calendar to help you answer the following questions:

1 What day of the week is 14th January 1988?

2 What day of the week is 1st March 1988?

3 What day of the week is 25th June 1988?

4 What day of the week is 31st August 1988?

5 What day of the week is 15th December 1988?

6 What date is the first Sunday in April?

7 What date is the last Monday in May?

8 What date is the third Thursday in November?

9 Which months in 1988 have five Sundays in them?

10 Is there a 'Friday 13th' in 1988? If so, in which month or months?

11 What date comes five days after February 26th?

12 What date comes twelve days after May 14th?

13 What date comes three *weeks* after July 4th?

14 What date comes three days *before* October 2nd?

15 What date comes six weeks before December 25th?

The Christian Church has its own special dates, many of them based on Easter. Easter Day 1988 was on Sunday 3rd April.

16 What date is Good Friday (two days before Easter)?

17 What date is Ascension Day (39 days after Easter)?

18 What date is Ash Wednesday ($6\frac{1}{2}$ weeks before Easter)?

B _____ Some dates have a special name and meaning – for instance, 25th December is Christmas Day. What is special about each of these dates?

1 1st January

2 14th February

3 1st March

4 1st April

5 23rd April

6 24th June

7 31st October

8 5th November

9 30th November

10 31st December

C _____ Some events do not always happen on the same date, but do always take place at roughly the same time of year. Say what is special about each of the following events, and when you would expect them to happen.

1 Burns' Night
2 Shrove Tuesday
3 Budget Day
4 Boat Race Day
5 Cup Final Day
6 Trooping the Colour
7 Wimbledon Fortnight
8 Harvest Festival
9 Yom Kippur
10 Diwali

D _____ Make a list (for your own use) of the important dates for your own family and friends. You might include all your friends' and relations' birthdays and wedding anniversaries, and a few other special dates like St Valentine's Day and Mothering Sunday. Use this list to help you remember to send cards or presents at the right times.

E _____ A small desk calendar is made from two cubes. Each cube has one figure on each face and by arranging the two cubes in a small frame they can be made to show any day of the month from 01 to 31.

● What six figures are on each cube? (There is more than one possible answer.)

F _____ In a class of 30, do you think it is likely that there will be two people who have the same birthday? Discuss this in your class and then check to see. Find out about other classes. (In fact the odds are about $2\frac{1}{2}$ to 1 that there will be two people with the same birthday in any group of 30.)

Unit 34 Speed

A car is travelling at 50 miles per hour (50 mph).
In 1 hour at this speed it goes 50 miles.
In 2 hours at this speed it goes $2 \times 50 = 100$ miles.
In 3 hours at this speed it goes $3 \times 50 = 150$ miles.
Of course the car might travel for less than 1 hour. In half an hour it will go 25 miles, but the speed is still 50 mph.

On the continent, car speeds are measured in kilometres per hour (km/h). A parachutist will measure his fall in feet per second (fps) or metres per second. Almost all speeds are in the form of **distance per time**, but ships use **knots** and supersonic aircraft use **Mach numbers**.

$$\text{Speed} = \frac{\text{distance}}{\text{time}}$$

If you know any two out of speed, distance moved and time taken, you can work out the third, as shown in the examples **a**, **b** and **c**.

a A cyclist travels at 15 km/h. How far can she go in $2\frac{1}{2}$ hours?
In 1 hour she goes 15 km, so in $2\frac{1}{2}$ hours she goes $2\frac{1}{2} \times 15 = 37\frac{1}{2}$ km.

b An aeroplane travels at 400 mph. How long does it take to fly 2400 miles? It travels 400 miles each hour, so for 2400 miles it takes $2400 \div 400 = 6$ hours.

c A parachutist falls 300 m in 60 s. What is his speed?
He falls 300 m in 60 seconds
so $\frac{300}{60} = 5$ m in 1 second $=$ m/s.

A

Use the examples to help you to do these.

1 A car drives along the motorway at 70 mph; how far will it go in 2 hours at this speed?

2 How far will the same car go in $2\frac{1}{2}$ hours?

3 A ship sails at 40 km/h; how long will it take to complete a 200 km journey?

4 How long will it take to go 300 km?

5 A hiker travels 9 miles in 3 hours; what is her average speed?

6 How far could this hiker travel in 4 hour at this speed?

7 How far could she travel in 24 hours? (Is this a sensible answer?)

8 Louella runs a marathon and covers 26 miles in 3 hours; what is her average speed?

9 Roger runs a mile in 4 minutes; what is his average speed in miles per hour?

10 Who is running faster, Louella or Roger? Who is the better runner?

11 A high-speed train runs at 120 mph; how far does it go in 1 minute?

12 How long does the train take to go 300 miles without stopping?

13 Sound travels at about 330 m/s (metres per second). How long would it take for the sound of an explosion to travel 1 km?

14 In a particular thunderstorm, it takes 5 seconds for the sound of the thunderclap to reach me; how far away is the storm?

15 A stone falls down a 64-foot deep well in 2 seconds. What is its average speed?

B

Use the same methods to tackle these questions, which are about other kinds of rates.

1 A baby puts on weight at a rate of 4 ounces per week. How much weight will it put on in 4 weeks?

2 How long will the baby take to put on 3 pounds?

3 A car uses petrol at a rate of 35 miles per gallon. How far will it go on 8 gallons of petrol?

4 How much petrol would the car use on a journey of 420 miles?

5 A tap produces water at a rate of 10 litres per minute; how long would it take to fill a 5-litre bucket?

Finding average speeds

A cyclist goes 30 km at 10 km/h, then 40 km at 20 km/h. What is his average speed?

The only way to solve a problem like this is to find the total distance and the total time.

You can work out that the cyclist took 3 hours for the first part of the journey and 2 hours for the second part of the journey.

distance = 30 + 40 = 70 km

time = 3 + 2 = 5 hours

so average speed = 70 ÷ 5 = 14 km/h

C _____ Work out the average speed for the whole journey if a car goes for 1 hour at 60 miles per hour, then for 2 hours at 70 miles per hour, and finally for 1 hour more at 40 miles per hour.

D _____ A hiker went for a walk one day, and the graph on page 129 shows how far he had travelled after various periods of time. Use it to help you answer the following questions.

1   How far had he gone after 3 hours?
2   How far had he gone after $\frac{1}{2}$ hour?
3   At what time was he 7 km from where he started?
4   How far from the start is the place where he stopped for a rest?
5   How long did he stop there?
6   Did he eat anything while he was resting?
7   What was his average speed for the first part of the walk (i.e. before the rest)?
8   What was his average speed for the second part of the walk (after the rest)?
9   How far did he walk altogether?
10   What was his average speed for the whole journey?

Distance (km)

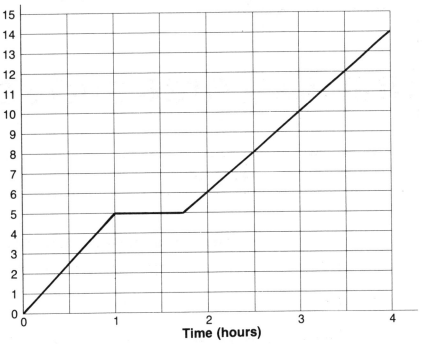

Time (hours)

E _____ A train 100 metres long is travelling along the line at 25 metres per second, and is approaching a tunnel 200 metres long.

● How long will it take the train to pass completely through the tunnel, from the time the engine goes in to the time the last carriage comes out at the far end?

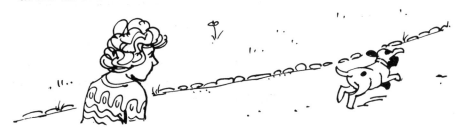

F _____ A girl is walking home with her dog at a steady speed of 3 mph. When they are just 1 mile from home, the dog leaves its mistress and runs on ahead at 12 mph. When it reaches home, it finds the door shut, so it immediately turns round and runs back (still at 12 mph) to meet the girl, who has kept walking at 3 mph all the time. As soon as the dog reaches her, it turns round again and runs home, and it keeps running to and fro between home and mistress until the girl too reaches home and opens the door.

● How far altogether has the dog run since it first left its mistress?

 Area and Perimeter

Areas

The **size** of a **surface** (flat or curved) is measured by its **area**.

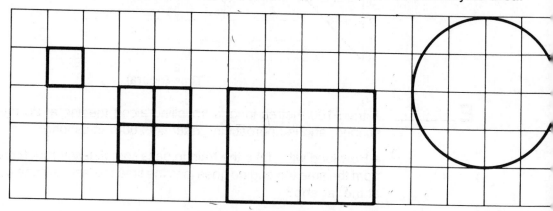

The first square has sides of 1 cm. It has an area of 1 square cm (1 cm²). The second square can be divided up into 4 squares with sides of 1 cm, so its area is 4 cm².

The oblong (rectangle) and circle are different shapes, but they have the same area. Check this by counting the squares.

A _____ Use the counting squares method to find the areas of the shapes in this exercise. Often you find that only parts of squares are covered. If so you can simply **count** the square when the shape covers **more** than half and **not count** the square if it covers **less** than half.

The square cm is a very small unit. To measure the area of a country we use bigger units such as **square kilometres** or **square miles**. Farmland areas are often measured in **acres**, or **hectares**. (A hectare is about the size of a football pitch. An acre is about half that.)

In the home, you are more likely to measure areas in **square feet** (ft² or **square metres** (m²).

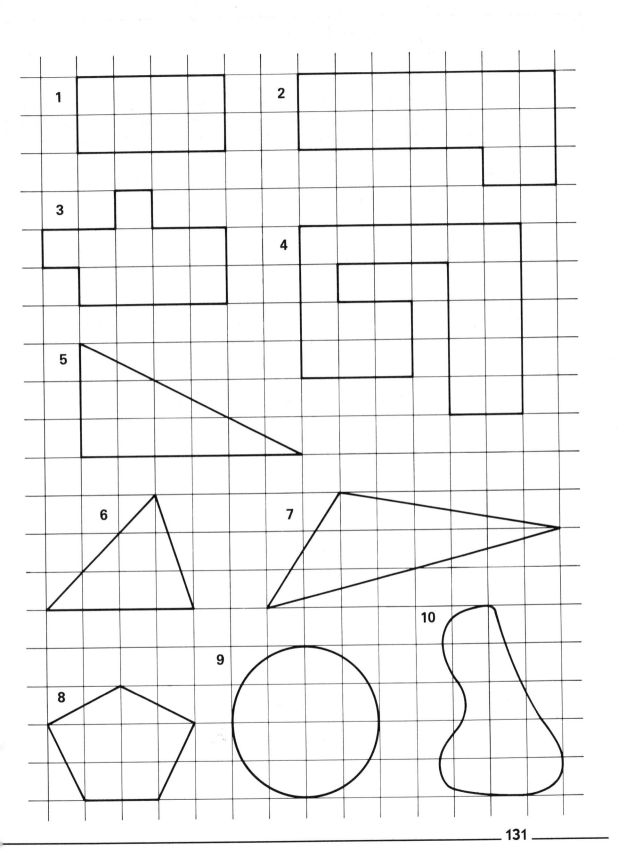

Perimeters

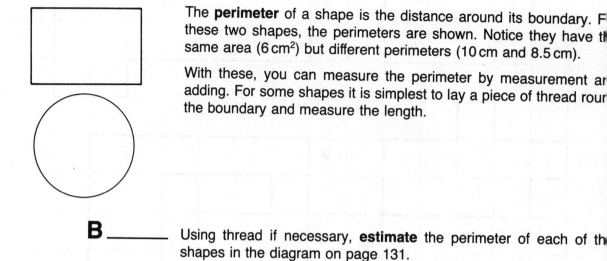

The **perimeter** of a shape is the distance around its boundary. F these two shapes, the perimeters are shown. Notice they have th same area (6 cm²) but different perimeters (10 cm and 8.5 cm).

With these, you can measure the perimeter by measurement an adding. For some shapes it is simplest to lay a piece of thread roun the boundary and measure the length.

B ———— Using thread if necessary, **estimate** the perimeter of each of th shapes in the diagram on page 131.

C ———— Of all the possible shapes with a perimeter of 20 cm, what shape d you think would have the largest area? Discuss your answer.

Unit 36 Rectangles

These diagrams all show rectangles. Check that they are all: **flat** with **4 straight sides** and **4 right angles** at the corners.

Use a ruler to check that: **opposite sides** are the **same length**.

Finding the perimeter and area of a rectangle

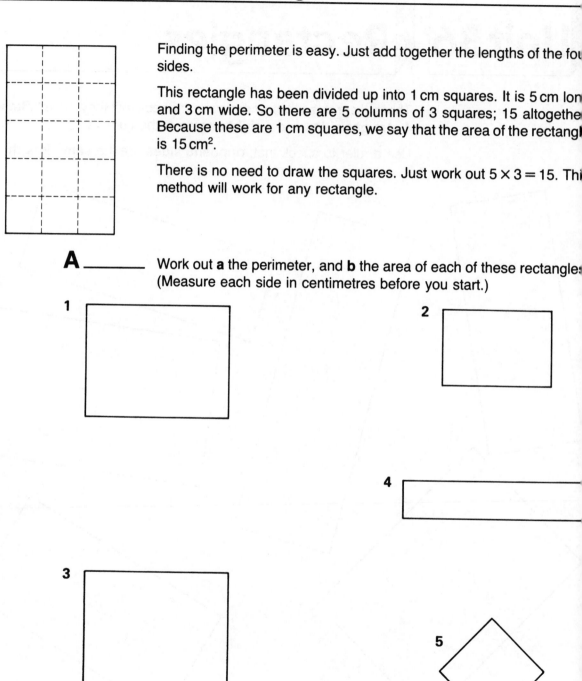

Finding the perimeter is easy. Just add together the lengths of the four sides.

This rectangle has been divided up into 1 cm squares. It is 5 cm long and 3 cm wide. So there are 5 columns of 3 squares; 15 altogether. Because these are 1 cm squares, we say that the area of the rectangle is 15 cm².

There is no need to draw the squares. Just work out 5 × 3 = 15. This method will work for any rectangle.

A ——— Work out **a** the perimeter, and **b** the area of each of these rectangles. (Measure each side in centimetres before you start.)

1

2

4

3

5

Work out **a** the perimeter, and **b** the area of each of the rectangles shown below. (These drawings are *not* to scale, but you should use the measurements given.)

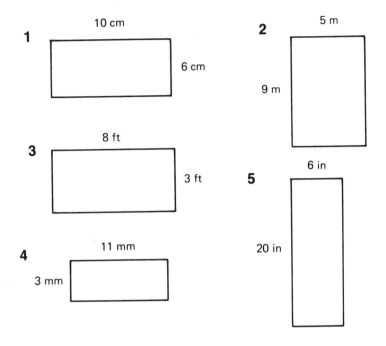

1 10 cm 6 cm

2 5 m 9 m

3 8 ft 3 ft

5 6 in 20 in

4 11 mm 3 mm

We can use similar methods in working out the areas and perimeters of more complicated shapes, such as the one here.

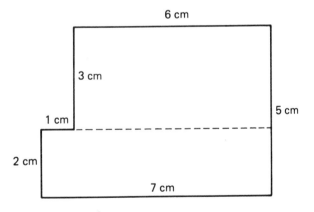

6 cm

3 cm

1 cm

5 cm

2 cm

7 cm

It is easy to find the **perimeter** by adding together the lengths of all the sides: $6+5+7+2+1+3=24$, so the perimeter is 24 cm.

To find the **area** of the shape we first divide it into two rectangles by drawing a dashed line.

The top oblong is 6 cm long and 3 cm wide, so its area is 18 cm²; the bottom oblong measures 7 cm by 2 cm and so has an area of 14 cm². Thus the area of the whole shape is $18+14=32$ cm².

C _____ Work out **a** the perimeter, and **b** the area of each of the shapes shown below. These are *not* drawn to scale; use the measurements given.

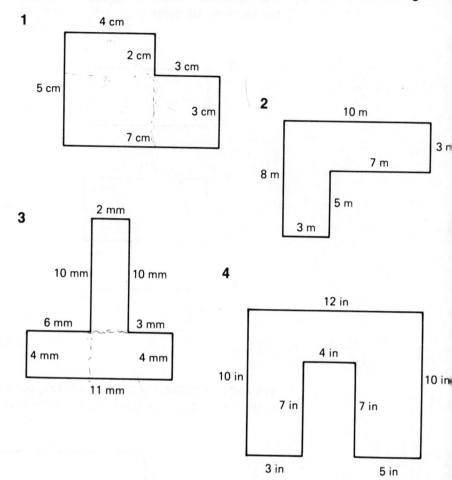

1

4 cm
2 cm
3 cm
5 cm
3 cm
3 cm
7 cm

2

10 m
3 m
7 m
8 m
5 m
3 m

3

2 mm
10 mm
10 mm
6 mm
3 mm
4 mm
4 mm
11 mm

4

12 in
4 in
10 in
10 in
7 in
7 in
3 in
5 in

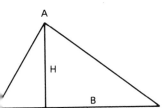

A triangle is a flat shape with 3 straight sides. In each of the triangles shown:

A is the **apex**.

B is the **base**.

H is the **height** of the triangle.

Finding the perimeter and area of a triangle

Finding the perimeter is easy. Just measure the lengths of the 3 sides and add them together.

Finding the area is not quite so easy. So here is an experiment to help you to understand the method.

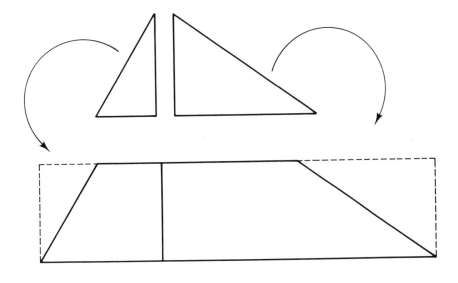

Take a sheet of sticky paper and draw a triangle like this but bigger
Measure **B** and **H** and write down the values. Now fold down the apex
of your triangle so it touches **B**. Crease the paper, unfold it and cut
along the crease. Also cut along the line H in the small top triangle. You
now have 3 pieces.

Stick the big piece down. Fit in the two small ones as shown.

Now you have a rectangle made up of the parts of the triangle. The
shape is different but **the area is the same**.

Area of rectangle = base × height

But base of rectangle = base of triangle

Height of rectangle = $\frac{1}{2}$ height of triangle

So area of triangle = base × $\frac{1}{2}$ height
$$= B × \tfrac{1}{2}H$$

The results can be applied to any triangle. The area of this one is 10 × $\frac{1}{2}$
of 8 = 40 cm².

A ———— Work out **a** the perimeter, and **b** the area of each of the triangles
shown below. (These are *not* drawn to scale; use the measurements
given.)

1

5 cm, 8 cm, (4 cm), 10 cm

2

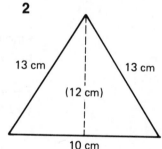

13 cm, 13 cm, (12 cm), 10 cm

3

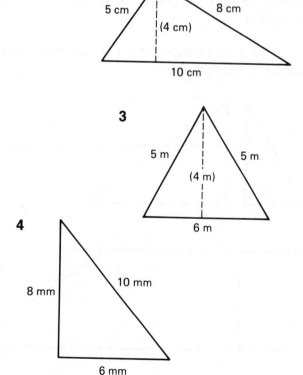

5 m, 5 m, (4 m), 6 m

4

8 mm, 10 mm, 6 mm

5

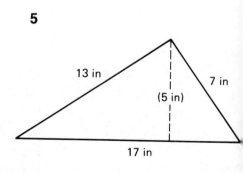

13 in, 7 in, (5 in), 17 in

B _____ The two triangles shown here are quite different in shape, but both have a perimeter of 24 cm. There are other triangles (with sides which are whole numbers of centimetres) which also have a perimeter of 24 cm.

List all such possible triangles. Which of them would you guess has the largest **area**?

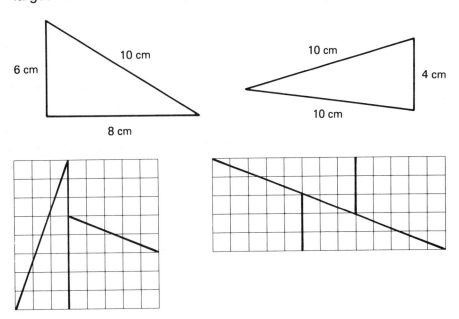

C _____ Draw a 8 cm by 8 cm square on a piece of squared paper. Mark this square as shown in the left-hand diagram above and cut out the pieces. Now rearrange them to make a 13 cm by 5 cm oblong, as shown in the right-hand diagram.

The square had an area of 64 cm² (8 × 8), and the oblong has an area of 65 cm² (13 × 5). Where has the extra square centimetre come from?

D _____ Imagine that you are the manager of a steelworks. You have three squares of sheet steel, each measuring 10 feet by 10 feet. A customer comes to you and asks you to cut from these:

    4 rectangles each measuring 7 feet by 3 feet

    6 rectangles each measuring 6 feet by 4 feet

    3 rectangles each measuring 5 feet by 3 feet

Draw sketches to show how you would cut each of the three large squares to produce these pieces. (It's not as easy as it sounds!)

Unit 38 Circles

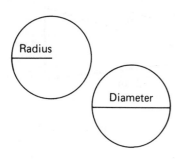

Draw a circle using a pair of compasses.

You have drawn a line that is always the same distance (the **radius** *r* from a fixed point (the centre).

The **diameter** (*d*) is the distance from one edge of the circle to the other through the centre. You can see straight away that $d = 2r$.

Circumference of a circle

The perimeter of a circle is called its **circumference**.

Draw a circle of radius 5 cm (diameter 10 cm). Use some thread to measure the circumference. If you are careful, you will find that the circumference is just over 30 cm. Roughly, it is all right to say that

$$\text{circumference} = 3 \times \text{diameter}.$$

If you measured the circles accurately you would find that

$$\text{circumference} = 3 \cdot 14159265 \ldots \times \text{diameter}.$$

$\pi = 3 \cdot 141592653589 \ldots$
We use $\pi = 3$
or $\quad \pi = 3 \cdot 14$
or $\quad \pi = 3\frac{1}{7} = \frac{22}{7}$

The number used here is π (pronounced 'pie'). It goes on forever without repeating itself, but normally we don't need to be this accurate. Use 'roughly 3', or 3.14, or $3\frac{1}{7}$ ($\frac{22}{7}$). In a test or examination, you will always be told which to use.

These examples show how you would find the circumference of a circle of diameter 9 cm. Of course, if your calculator has a π button, the calculation is made easier.

> **a** circumference $= 3 \times 9 = 27$ approx
>
> or **b** circumference $= 3 \cdot 14 \times 9 = 28 \cdot 26$
> $= 28$ cm to the nearest whole number.
>
> or **c** circumference $= 3\frac{1}{7} \times 9 = 27 + \frac{9}{7} = 28\frac{2}{7}$
> $= 28$ cm to the nearest whole number.

Ask your teacher whether you can use a calculator for the following questions.

A

In questions 1 to 5 use 3·14 as your value for π.

1 Work out the circumference of a circle with diameter 5 cm.

2 Work out the circumference of a circle with diameter 12 feet.

3 A circle has a diameter of 4 m; what is its circumference?

4 Find the circumference of a circle whose diameter is 3 yards.

5 Find the circumference of a circle with a *radius* of 3 km. (*Hint:* what is the connection between the radius and the diameter?)

6 Using $3\frac{1}{7}$ for π, find the circumference of a circle with a diameter of 14 cm.

7 using $3\frac{1}{7}$ again, find the circumference of a circle whose diameter is 10 m.

8 Using 'just over 3' for π, estimate the circumference of a circle with a diameter of 32 in.

9 Using the value 3·14159265 for π, work out the circumference of a circle with diameter 100 km.

10 Convert your answer from question 9 into millimetres. Could you really measure the circumference this accurately?

Area of a circle

We also use π to find the area of a circle.

This circle has a radius of 1 inch. Its area is $3·14159\ldots$ in^2 = π in^2.

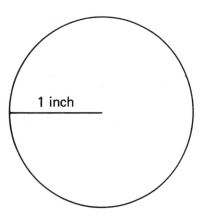

1 inch

A circle of radius 2 inches has an area of $4 \times 3.14159\ldots \text{in}^2 = 4\pi \text{ in}^2$

A circle of radius 3 inches has an area of $9 \times 3.14159\ldots \text{in}^2 = 9\pi \text{ in}^2$

In other words, **area = (radius)² × π.**

B _____ In questions 1 to 5, use 3·14 as your value for π.

1 Find the area of a circle with radius 4 cm.

2 Work out the area of a circle whose radius is 10 cm.

3 A circle has radius 2 m; calculate its area.

4 Calculate the area of a circle with radius 5 miles.

5 Calculate the area of a circle whose *diameter* is 12 km.

6 Using $3\frac{1}{7}$ for π, find the area of a circle with radius 14 feet.

7 Using 'just over 3', estimate the area of a circle with radius 100 km.

8 Using $\pi = 3.14159265$, work out the area of a circle whose radius is 3 cm; then round off your answer to two decimal places.

9 The ancient Egyptians used a different rule altogether. They said 'Take one-ninth of the diameter away, and square the result.' Use the Egyptian method to find the area of a circle with a diameter of 18 cubits.

10 Now use our modern method (with π as 3·14) to find the area of a circle with a *diameter* (be careful) of 18 cubits. (The answer will be in square cubits.)

C _____ Use the methods of the last three Units to help you answer these questions.

1 A bicycle wheel has a diameter of 28 inches. How far will it travel in one complete revolution? (Take π as $3\frac{1}{7}$.)

2 A hockey pitch is 100 yards long and 60 yards wide. If Joyce runs all the way round it, how far does she run altogether?

3 An ornamental fish-pond is circular, with a diameter of 8 feet. What is its surface area? (Take π to be 3·14.)

4 A triangular sailing course consists of three 'legs'. The contestants first sail 14 km north, then 48 km west, and finally 50 km back to the start. How far do they sail altogether?

5 Work out (a) the perimeter, and (b) the area of each of the shapes shown below. (Use $\pi = 3.14$ when necessary.)

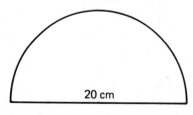

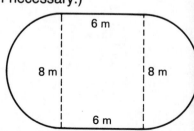

Unit 39 Volumes of Solids

The volume of a solid is the amount of space it takes up.

To measure volume, we try to find out how many unit cubes the object might be divided into.

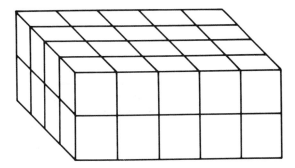

This cuboid is 5 cm long, 4 cm wide and 2 cm high. Divided into centimetre cubes we see that there are two layers of cubes, each containing 4 rows of 5 cubes each.

In each layer, there are 4×5 cubes $= 20$ cubes.

Two layers give 2×20 cubes $= 40$ cubes.
   So volume $= 40$ cm^3.
You can see this is just $4 \times 5 \times 2$ or

length $\times$ **breadth** $\times$ **height**.

In this second cuboid, the volume is $12 \times 9 \times 4 = 432$.

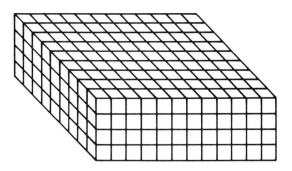

A _____ Work out the volume of each of these cuboids.

1 A cuboid 10 cm long, 5 cm wide and 3 cm high.
2 A cuboid 8 cm long, 2 cm wide and 2 cm deep.
3 A cuboid 4 feet long, 2 feet wide and 3 feet deep.
4 A cuboid 6 metres long, 5 metres wide and 1 metre deep.
5 A cuboid 100 cm long, 3 cm wide and 2 cm high.

6
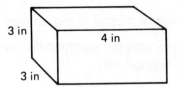

7
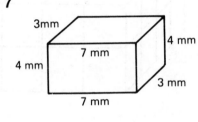

8
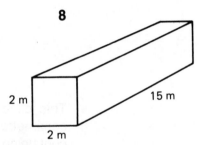

9
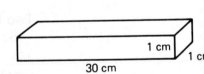

10
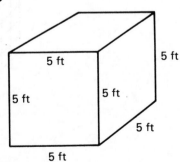

Nets

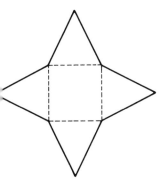

If we are going to make a cuboid (or any other solid shape) from cardboard, we normally begin with a pattern called a **net**.

The pattern here, for example, could be cut along the solid lines and folded along the dotted ones to make a pyramid, so we say that it is the net of a pyramid.

B _____ On cardboard, carefully draw the net of a cube with edges 5 cm long. Test your answer by cutting it out and folding it into shape; if it fits, stick it together with adhesive tape.

C _____ The diagram below (not to scale) shows the net of a cuboid.

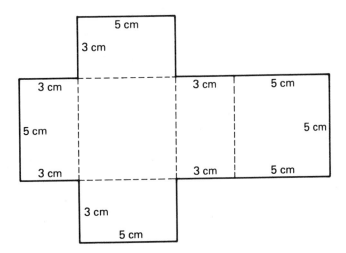

1   What area of cardboard does it use?

2   When the cuboid is made up, what is its volume?

Unit 40 | Ratios

Some simple ratios

To make a shade of orange paint you mix 2 tins of yellow paint with tin of red paint. To make more orange paint of the same colour, yo could:

mix 4 tins of yellow with 2 tins of red, or

mix 20 tins of yellow with 10 tins of red.

In each case the number of yellow tins is twice the number of red tin So the **ratio** of yellow:red is two to one or 2:1.

Suppose you make green paint by mixing blue and yellow in the rat 2:3. If you start with 12 tins of blue, how much yellow will be needed

12 tins of blue contain six lots of 2 tins.

So the yellow needed is six lots of 3 tins = 18 tins.

A —————— We can make grey paint by mixing black and white paint in the ratio 2: (that is, two tins of black with five tins of white).

1 If we start with 6 tins of black paint, how much white paint do w need?

2 If we start with 10 tins of black, how much white do we need?

3 If we start with just one tin of black paint, how much white paint d we need?

4 If we start with 20 tins of *white* paint, how much black do we need

5 If we want to make 14 litres of grey paint altogether, how muc black and how much white should we mix?

B _____ Cumberland Rum Butter is made by mixing sugar and butter in the ratio 2:1 by weight (that is, 2 ounces of sugar for every ounce of butter) and then adding flavourings.

1 If we start with 8 ounces of butter, how much sugar do we need?

2 If we start with 8 ounces of sugar, how much butter do we need?

3 If we start with 500 grams of butter, how much sugar should we add?

4 How much sugar should be mixed with 1 lb butter?

5 If we start with 800 grams of sugar, how much butter needs to be added?

Equivalent ratios

$1:2 = 2:4 = 10:20$

As we saw with the tins of paint, all these are the same (just as $\frac{1}{2} = \frac{2}{4} = \frac{10}{20}$).

With fractions we reduce to the lowest terms. Ratios are also reduced to the lowest terms.

So with 6:9, we divide each side by 3 and write $6:9 = 2:3$.

C _____ Write each of these ratios in their lowest terms.

1 5:10

2 10:5

3 6:8

4 9:3

5 6:12

6 4:10

7 8:12

8 20:12

9 16:24

10 15:10

We sometimes have occasion to use a ratio containing more than two terms. For example, a person who likes sweet white coffee might decide to mix instant coffee powder, sugar, and dried milk together in the ratio 1:3:4. That might mean one spoonful of coffee, three of sugar and four of milk; or if she is making a jugful it might need three spoonfuls of coffee, nine of sugar and twelve of milk. As long as we remember the order of the items, ratios like this are really no more difficult than those in Unit 41.

D

A rich uncle always gives his nephews and nieces cash birthday presents in the same ratio as their ages. This year, Anne is 12, Bryn is 10 and Charles is 5, so the ratio this year is 12:10:5.

1 If Charles gets £5 this year, how much does Bryn get?
2 If Bryn gets £5 this year, how much does Anne get?
3 If Anne gets £60 this year, how much does Bryn get?
4 If Charles gets £20 this year, how much does Anne get?
5 Next year, of course, all three children will be a year older. What will the ratio of values be *next* year?
6 If Charles gets £30 this year, how much does Bryn get?
7 If Charles gets £30 *next* year, how much does Bryn get then?
8 If Charles gets £30 next year, how much will Anne get then?
9 If Charles got £20 *last* year, how much did Anne get?
10 This year the children's uncle sent them £270, which he said they were to share between them in the ratio of their ages. How much of the money should each child have got?

More Ratios and Scales

Unequal sharing

60 sweets shared in ratio 1:2

If David and Goliath share 60 sweets equally (ratio 1:1) each gets 30. Suppose Goliath, who is twice as big, wants twice as many. Then we need 3 shares – one for David, two for Goliath.

Each share is $60 \div 3$ = 20 sweets
So David gets 1 lot of 20 = 20 sweets
And Goliath gets 2 lots of 20 = 40 sweets

270 shared in ratio 12:10:5

Say there are $12 + 10 + 5 = 27$ shares. Each share is worth $£270 \div 27 = £10$. So the three parts are:

12 shares $= £10 \times 12 = £120$
10 shares $= £10 \times 10 = £100$
5 shares $= £10 \times 5 = £50$

(As a check, $£120 + £100 + £50 = £270$, which is right.)

A

1 Share 45 sweets between Martha and Mary so that Mary has twice as many as Martha.

2 Share 100 assorted marbles between Peter and Paul in the ratio 2:3.

3 Share £50 between Patrick and Andrew in the ratio of their ages, which are 7 and 3 respectively.

4 The three angles of a triangle are in the ratio 2:3:4, and add up to 180°. How many degrees is the smallest angle?

5 A garden compost is made from fertilizer and loam in the ratio 1:9 by weight. How much of each will there be in 20 kg of compost?

6 Olga, Masha and Irina have 350 roubles to share between them i
the ratio 6:5:3. How much should each sister get?

7 Divide 100 gallons of treacle between Elsie, Lacie and Tillie in th
ratio 2:5:3.

8 A triangle with sides in the ratio 5:12:13 has a perimeter c
45 cm. How long is the longest side?

9 Share £6000 profit between five business partners in the rati
7:5:3:3:2.

10 Divide 42 biscuits between Faisal, Fuad and Fatima so that Faisa
gets twice as many as Fuad and Fatima gets twice as many a
Faisal.

Scale models

This OO gauge model railway is made on a scale of $\frac{1}{75}$, so that the rati
of model length: real life length is 1:75. The table shows a compariso
of the model with the real thing.

Model		Real thing
carriage length	20 cm	length 75 × 20 = 1500 cm = 15 m
	8 in	75 × 8 = 600 in = 50 ft
signal	15 cm high	75 × 15 = 1125 cm = 11.25 m
	6 in high	75 × 6 = 450 in = $37\frac{1}{2}$ ft

B _____ A model warship is made on a scale of 1:600.

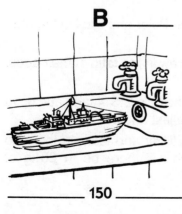

1 If the model is 25 cm long, how long is the real ship?

2 If the model has a beam (width) of 5 cm, what is the real ship'
beam?

3 If the real ship is 48 m high (4800 cm), how high is the model?

4 If the guns of the real ship are 6 m long, how long are the model'
guns?

5 If the model has 2 funnels, how many funnels has the real ship'

C _____ A dolls' house is made on a scale of 1:20.

1 If the model is 2 feet high, how high would the real house be?

2 If the model hall is 9 inches long, how long would the real hall be?

3 The real lounge measures 180 inches by 140 inches; what would be the dimensions of the model lounge?

4 The real house has a staircase with 20 steps; how many steps will there be on the model staircase?

5 The model house weighs 5 pounds; what would be the weight of the real house if it were made of the same materials? Does that sound right?

Maps and plans

A map or plan is like a model in some ways. But the scale is usually shown in a different way.

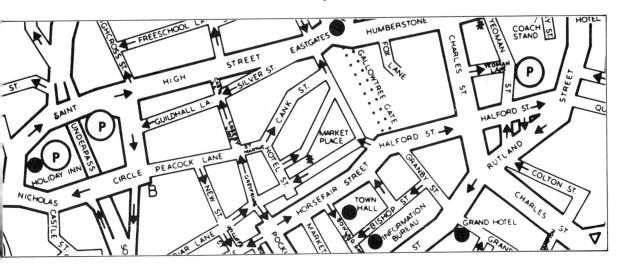

This street plan shows part of a town. The **scale** is 6 inches to 1 mile. To write this as a ratio, the units must be the same.

1 mile = 63 360 inches
so the ratio is 6:63 360
or 1:10 560.

Landranger Ordnance Survey maps are on a scale of 2 cm to 1 km.
1 km = 100 000 cm, so the ratio is 2:100 000
or 1:50 000

D _____ Write each of these scales as a ratio in its **lowest terms**.

1 2 inches to 1 mile
2 1 inch to 2 miles
3 10 cm to 1 km
4 1 cm to 4 km
5 1 cm to 2 m
6 5 cm to 1 m
7 5 cm to 1 km
8 1 inch to 6 feet
9 1 inch to 50 feet
10 10 inches to 1 mile

In this small map, the scale is 2 cm to 1 km. The villages are 6 cm apart on the map. 2 cm represents 1 km, so the villages are really 3 km apart.

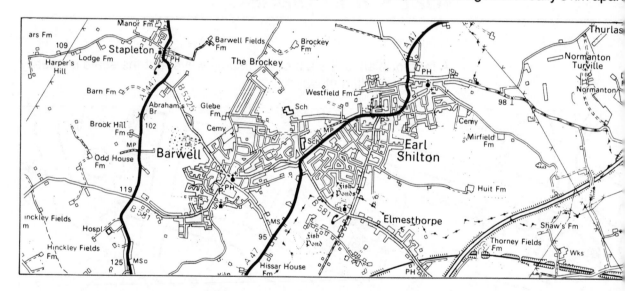

E _____ A *Landranger* map is drawn on a scale of 2 cm to 1 km.

1 Two hilltops are 20 cm apart on the map; how far apart are the real hills?

2 A road on the map is 16 cm long; how long is the real road?

3 A lake is really 15 km long; how long will the lake on the map be?

4 Two railway stations are 13 cm apart on the map; how far apart are the real stations?

5 Two motorway junctions are actually 25 km apart; how far apart will they be marked on the map?

F _____ A road map is drawn on a scale of 1 inch to 8 miles.

1 Two towns on this map are 3 inches apart; how far apart are they really?

2 A canal shown on the map is drawn 7 inches long; how long is the actual canal?

3 The distance from Liverpool to Manchester is 32 miles; how far apart will these two towns be marked on the map?

4 A motorway shown on the map is $\frac{1}{8}$ inch wide; how wide is the real motorway?

5 Does this answer make sense? If not, what does it show about the way maps are drawn?

Unit 42 | Proportion

Direct proportion

> If 6 oranges cost 54p, how much will 8 oranges cost?

The easy way to solve this problem is to find out how much 1 orange costs:

$$1 \text{ orange costs } 54 \div 6 = 9p$$

then find the cost of eight oranges:

$$8 \times 9p = 72p$$

> 300 drawing pins weigh 60 g. What do 500 drawing pins weigh?

Here is another example with larger numbers. Instead of finding the weight of 1 drawing pin, find the weight of 100 ('1 packet').

$$300 \text{ pins weigh } 60g$$
$$100 \text{ pins weigh } 60 \div 3 = 20g$$
$$\text{so } 500 \text{ pins weigh } 5 \times 20 = 100g$$

A _____

Use the methods described above to answer the following questions

1 If 4 bananas cost 44p, what would 6 bananas cost?
2 If four lemons weigh 20 ounces, what do three lemons weigh?
3 In six days a ship can sail 2100 miles; how far can it go in ten days?
4 If two dozen eggs cost £1.66, what do three dozen eggs cost?
5 If fifteen carpet tiles will cover 2·4 m², what area will 25 tiles cover?
6 If 28 rows of knitting measure 7 cm, what do 36 rows measure?
7 A clock gains 3 minutes in 18 hours; how many minutes will it gain in a week (168 hours)?

8 A block of lead with a volume of 30 cm³ weighs 342 g; what would be the weight of a block with volume 40 cm³?

9 If 24 ounces of pastry will make 36 mince pies, how many mince pies will 36 ounces of pastry make?

10 In five hours I can cycle 60 miles; how far can I cycle in three hours at the same speed?

Inverse proportion

If you buy more **oranges**, they **cost** more. If you have more **drawing pins**, they **weigh** more. If you travel more **miles**, it takes a longer **time**.

In all these cases, when one quantity gets bigger, so does the other.

Now suppose that one man can build a wall in 20 days. With two men it should take less time; 10 days. Four men could do the job in only 5 days. This is an example of **inverse proportion**, that is, when an increase in one quantity causes a decrease in another.

Here is a simple problem:

10 workers build a wall in 3 weeks. How long will 15 workers take?

Think about 1 worker alone. It took 3 weeks for 10 workers, so 1 worker will need 10 × 3 = 30 weeks. Now 15 men share this work, so they will need 30 ÷ 15 = 2 weeks.

B _____ Use inverse proportion to solve these problems.

1 Five labourers can dig a mile-long ditch in eight days. How long would just four labourers take to dig a similar ditch?

2 In a lifeboat there is enough water to last ten people for fourteen days. If there are only seven people in the boat, how long will the water last?

3 If four pumps are used, a swimming bath can be filled in nine hours. How long will it take to fill the bath if six pumps are used?

4 At a speed of 30 knots, a certain journey takes a ship 10 days. How long would the journey take if the speed were reduced to 20 knots?

5 If I spend £2.25 per day while I am on holiday, my holiday money will last for seven days. How long will it last if I spend only £1.75 per day?

Using common sense

a It takes $4\frac{1}{2}$ minutes to boil an egg. How long will it take to boil 3 eggs?

b Radio 2 has a wavelength of 330 m. What is the wavelength of Radio 4?

The answers are:

a $4\frac{1}{2}$ minutes – all the eggs can cook in the same saucepan.

b 1500 m. The wavelength has nothing to do with the number of the station.

In these examples, there is no connection between the two things mentioned. Neither direct nor inverse proportion can be applied. **Watch out for situations like these.**

c _____ Before you try to answer each of the next questions, ask yourself:

Does doubling one thing double the other (direct proportion)?
Does doubling one thing halve the other (inverse proportion)?
Does doubling one thing have no effect on the other (no connection)?

1 Three dozen buttons cost 90p; what do five dozen buttons cost?

2 Five eggs weigh 150 g; what do nine eggs weigh?

3 Two men on a beach can see eight miles; how far can three men see?

4 Five loaves of bread will make 55 sandwiches; how many sandwiches will eight loaves make?

5 At age seven, a boy is four feet tall; how tall is he at age fourteen?

6 There are 12 feet in four yards; how many feet are there in five yards?

7 A barn contains enough hay to feed 30 cows for 8 weeks; for how long would this hay feed 40 cows?

8 Seven sergeants are responsible for 140 soldiers; if the squads are all the same size, for how many soldiers are ten sergeants responsible?

9 In five minutes a secretary can type 150 words; how many words can she type in twenty minutes?

10 In a shift, four fitters can lay 120 metres of pipe; how much pipe could six fitters lay in the same time?

11 At 3 o'clock the temperature is 24 °C; what is the temperature at 4 o'clock?

12 Three sacks of fertilizer will cover $1\frac{1}{2}$ acres; what area will five sacks cover?

13 At 30 mph a certain journey takes 4 hours; how long would it take at 20 mph?

14 In 30 minutes a gardener can mow 500 m² of lawn; how much lawn can she mow in 45 minutes?

15 In a race, horse number 2 has four legs; how many legs has horse number 3?

16 If five pears weigh 600 g, how many pears weigh 960 g altogether?

17 If five pounds is equivalent to seven dollars, how many dollars are equivalent to twenty pounds?

18 In my car, seven gallons of petrol will take me 245 miles; how far will five gallons take me?

19 Three singers can sing a song in six minutes; how long will five singers take to sing it?

20 A cow eats four bales of hay in twelve days; how long will five bales last?

D _____ Try these brain teasers!

1 Four years ago, John was 8 and his mother was 32, so she was four times as old as he was. Now, John is 12 and his mother is 36, so she is only three times as old as him. In how many more years will John's mother be just twice his age?

2 Hughie has twice as many marbles as Irene, but if he gives her ten of his marbles, she will then have three times as many as him. How many marbles has Hughie at the moment?

3 Thomas and Tracey are brother and sister. Thomas has twice as many sisters as he has brothers, but Tracey has the same number of sisters as brothers. How many boys, and how many girls, are there in their family?

4 Four children at the cinema are sitting in a row of four seats. David is sitting next to Doreen; Diana is sitting between David and Duncan. Who is at each end of the row?

5 Two Russians are walking across Red Square. The big Russian is the little Russian's father, but the little Russian is **not** the big Russian's son. How is that possible?